Data Orchestration
in Deep Learning Accelerators

Synthesis Lectures on Computer Architecture

Editors
Natalie Enright Jerger, *University of Toronto*
Margaret Martonosi, *Princeton University*

Founding Editor Emeritus
Mark D. Hill, *University of Wisconsin, Madison*

Synthesis Lectures on Computer Architecture publishes 50- to 100-page books on topics pertaining to the science and art of designing, analyzing, selecting, and interconnecting hardware components to create computers that meet functional, performance, and cost goals. The scope will largely follow the purview of premier computer architecture conferences, such as ISCA, HPCA, MICRO, and ASPLOS.

High Performance Datacenter Networks: Architectures, Algorithms, and Opportunities
Dennis Abts and John Kim
2011

Processor Microarchitecture: An Implementation Perspective
Antonio González, Fernando Latorre, and Grigorios Magklis
2010

Transactional Memory, Second Edition
Tim Harris, James Larus, and Ravi Rajwar
2010

Computer Architecture Performance Evaluation Methods
Lieven Eeckhout
2010

Introduction to Reconfigurable Supercomputing
Marco Lanzagorta, Stephen Bique, and Robert Rosenberg
2009

On-Chip Networks
Natalie Enright Jerger and Li-Shiuan Peh
2009

The Memory System: You Can't Avoid It, You Can't Ignore It, You Can't Fake It
Bruce Jacob
2009

Fault Tolerant Computer Architecture
Daniel J. Sorin
2009

The Datacenter as a Computer: An Introduction to the Design of Warehouse-Scale Machines
Luiz André Barroso and Urs Hölzle
2009

Computer Architecture Techniques for Power-Efficiency
Stefanos Kaxiras and Margaret Martonosi
2008

Chip Multiprocessor Architecture: Techniques to Improve Throughput and Latency
Kunle Olukotun, Lance Hammond, and James Laudon
2007

Data Orchestration in Deep Learning Accelerators

Tushar Krishna, Hyoukjun Kwon, Angshuman Parashar, Michael Pellauer, and Ananda Samajdar

ISBN: 978-3-031-00639-5 paperback
ISBN: 978-3-031-01767-4 ebook
ISBN: 978-3-031-00064-5 hardcover

DOI 10.1007/978-3-031-01767-4

A Publication in the Springer series
SYNTHESIS LECTURES ON ADVANCES IN AUTOMOTIVE TECHNOLOGY

Lecture #52
Series Editors: Natalie Enright Jerger, *University of Toronto*
                Margaret Martonosi, *Princeton University*
Founding Editor Emeritus: Mark D. Hill, *University of Wisconsin, Madison*
Series ISSN
Print 1935-3235   Electronic 1935-3243

Data Orchestration
in Deep Learning Accelerators

Tushar Krishna
Georgia Institute of Technology

Hyoukjun Kwon
Georgia Institute of Technology

Angshuman Parashar
NVIDIA

Michael Pellauer
NVIDIA

Ananda Samajdar
Georgia Institute of Technology

SYNTHESIS LECTURES ON COMPUTER ARCHITECTURE #52

ABSTRACT

This Synthesis Lecture focuses on techniques for efficient data orchestration within DNN accelerators. The End of Moore's Law, coupled with the increasing growth in deep learning and other AI applications has led to the emergence of custom Deep Neural Network (DNN) accelerators for energy-efficient inference on edge devices. Modern DNNs have millions of hyper parameters and involve billions of computations; this necessitates extensive data movement from memory to on-chip processing engines. It is well known that the cost of data movement today surpasses the cost of the actual computation; therefore, DNN accelerators require careful orchestration of data across on-chip compute, network, and memory elements to minimize the number of accesses to external DRAM. The book covers DNN dataflows, data reuse, buffer hierarchies, networks-on-chip, and automated design-space exploration. It concludes with data orchestration challenges with compressed and sparse DNNs and future trends. The target audience is students, engineers, and researchers interested in designing high-performance and low-energy accelerators for DNN inference.

KEYWORDS

artificial intelligence (AI), deep learning, deep neural networks (DNN), convolutional neural networks (CNN), general matrix multiplication (GEMM), hardware/software co-design, deep neural network scheduling (DNN scheduling), deep neural network mapping (DNN mapping), dataflow, data orchestration, spatial accelerators, architecture, hardware

Contents

Preface

The rise of Deep Learning (DL) as a disruptive application domain, coupled with the end of performance scaling of traditional CPUs, has led to new paradigm of computer architectures, namely specialized accelerators for DL. Tens of billions of dollars have been invested in this area—with innovative accelerators being developed not only by the usual chip vendors (e.g., Intel, NVIDIA) but also by traditionally software companies (e.g., Google, Microsoft), start-tups, and academia. The goal of this Synthesis Lecture is to dissect and describe the key building blocks and design flows common across DL inference accelerators. In particular, we focus on efficient mechanisms to manage data orchestration—i.e., systematically staging fine-grained data movement within an accelerator for performance and energy efficiency. The expected background for readers is an understanding of basic computer architecture. Our aim is to bootstrap students and designers into this rapidly growing and exciting domain.

In this book, we walk through the various mechanisms to enable efficient data orchestration within DNN inference accelerators—which could be deployed in the cloud or on edge platforms. Chapter 2 provides necessary background on Dataflow and Data Reuse. Chapter 3 presents the first key component of data orchestration—on-chip buffer organization. Chapter 4 describes the second key component of data orchestration—on-chip network architectures. Chapter 5 ties the dataflow, buffer hierarchies, and on-chip networks together into a systematic methodology for designing DNN accelerators, and it also presents a case study of current state-of-the-art accelerators. Chapter 6 delves into the design-space-exploration approaches for DNN accelerator design. Chapter 7 presents a preview of ongoing activity in architecting accelerators for sparse DNNs. Finally, Chapter 8 concludes with a brief discussion of DNN acceleration approaches that this book did not delve into and future research opportunities.

Tushar Krishna, Hyoukjun Kwon, Angshuman Parashar, Michael Pellauer,
and Ananda Samajdar
July 2020

Acknowledgments

We would like to start by thanking Natalie Enright Jerger (the current editor), Margaret Martonosi (the previous editor), and Michael Morgan (President and CEO of Morgan & Claypool Publishers) for inviting us to write this Synthesis Lecture and for their continuous feedback and encouragement throughout the process. A special thanks to Joel Emer for detailed technical discussions with us on accelerator design and related topics through the course of multiple years that have culminated in many of the technical idioms described in this book. In addition, many thanks to Vivienne Sze, Yu-Hsin Chen, and Joel Emer for their fundamental contributions to the topic of DNN acceleration via the Eyeriss chip, dataflow taxonomy, and pedagogical tutorials. The Synthesis Lecture titled *Efficient Processing of Deep Neural Networks* by Vivienne Sze, Yu-Hsin Chen, Tien-Ju Yang, and Joel Emer is a synergistic and complementary take on the topics discussed in this book. We also acknowledge our collaborators and co-authors across multiple papers who helped shape much of the content in this book: Prasanth Chatarasi, Eric Qin, and Vivek Sarkar (Georgia Tech); Christopher W. Fletcher and Kartik Hegde (University of Illinois Urbana-Champaign); Sophia Shao (UC-Berkeley); Jason Clemons, Neal Crago, Aamer Jaleel, Rangharajan Venkatesan, Brucek Khailany, Stephen W. Keckler, and William J. Dally (NVIDIA); Priyanka Raina (Stanford); Victor A. Ying and Anurag Mukkara (MIT); Minsoo Rhu (KAIST); Antonio Puglielli (Zendar); Dipankar Das and Sudarshan Srinivasan (Intel); and Paul Whatmough and Matthew Mattina (ARM). Last, but not the least, we appreciate the detailed feedback and suggestions from Christopher W. Fletcher (University of Illinois Urbana-Champaign) and Tony Nowatzki (UCLA) that were invaluable in improving the manuscript.

Tushar Krishna, Hyoukjun Kwon, Angshuman Parashar, Michael Pellauer,
and Ananda Samajdar
July 2020

CHAPTER 1

Introduction to Data Orchestration

We begin with a chapter that introduces the overarching theme of this Synthesis Lecture: hardware acceleration of Deep Learning inference. First, we provide a brief background on Deep Neural Networks (DNNs), which are the underlying computational mechanisms within Deep Learning applications. Our objective is not to go into the theory behind the structure and accuracy of DNNs (which readers can find in any modern textbook on Machine Learning or Deep Learning), but rather to identify the key computations within DNNs that are being considered for acceleration throughout the book. Next, we introduce DNN accelerators, describing the template architecture we consider in this book—namely a spatial array of processing elements connected via a network-on-chip to a custom buffer hierarchy. Finally, we coin the term "data orchestration" to refer to the systematic staging of data movement within the accelerator for data reuse and efficiency.

1.1 DEEP NEURAL NETWORKS (DNNS)

Neural networks are a rich class of algorithms that can be trained to approximate the behavior of complex mathematical functions. They are inspired by human brains and comprise of a large collection of *neurons* connected with *synapses*. Each neuron is connected with many other neurons and its output enhances or inhibits the actions of the connected neurons. These connections are called synapses, and each synapse has a *weight* associated with it. Modern neural networks are typically architected as a series of layers. For this reason they are called they are called "deep" neural networks or DNNs. The first (a.k.a. input) layer takes the external inputs (a.k.a. activations), followed by multiple internal (a.k.a. hidden) layers, followed by the final (a.k.a. output) layer that provides the final output for the function being approximated.

Figure 1.1a shows the general structure of DNNs. The neuron is the fundamental building block of DNNs. Its structure is shown in Figure 1.1b. It computes a weighted-sum to incorporate various features of inputs in different degrees to generate an output value, and applies a nonlinear activation function to the output value to generate the final output. The degree of consideration for each input feature depends on the weight on the corresponding synapse.

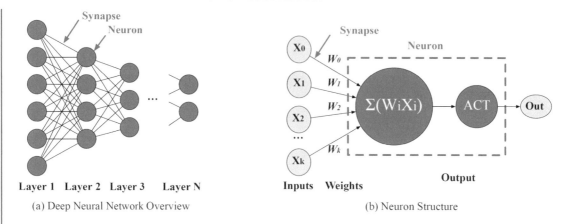

(a) Deep Neural Network Overview   (b) Neuron Structure

Figure 1.1: Structure of deep neural networks. Each synapse (i.e., connection) in (a) and (b) represents a weight. The connectivity can change depending on the type of DNN. The weight values are determined after training.

1.1.1   DNN TRAINING AND INFERENCE

DL today operates in two phases: training and inference. During training, a DNN is fed multiple samples of *labeled datasets*. Labeled datasets tag each data sample with the correct output, also called *ground truth*. For instance, a picture of a dog tagged with a label "dog." These data samples are streamed through the DNN from the input layer to the output layer; this is known as a *forward pass*. For each data sample, the error between the predicted output and the ground truth is computed and propogated through the DNN in reverse, from the output layer to the input layer. This is known as *backpropagation*. During backpropagation, the synaptic weights are adjusted with the aim of minimizing the gradient of the error with respect to each weight. This process is called *gradient descent*. However, since calculating the true error across all images would be prohibitively expensive, in practice an optimization called *stochastic* gradient descent is used to calculate an estimated gradient. In either case, because the ground truth must be known a priori, the process described above is referred to as *supervised* learning, as humans were involved in the labeling of the dataset.

The training loop is considered to be complete when the DNN predictions on input samples that have not been used for backpropagation (the *testing set*) exceeds the user's accuracy requirements. The trained DNN is then deployed for use by end applications. For legacy reasons, the trained DNN is often called a *model*, since the underlying function approximator need not be a neural network and can be some other machine learning (ML) approach as well.[1]

After deployment, the DNN makes inferences (i.e., predictions) on new data without any notion of ground truth. Only a forward pass through the layers is done during inference. In

---

[1]The discussion of other ML approaches is beyond the scope of this book.

general, weights are not updated dynamically in the field, except as the result of another training process. Therefore, the training and inference steps should be considered as separate workloads, with training generally being a superset of inference.

An important factor to take into account while deciding the hardware design for training and inference workloads is the emphasis on latency vs. throughput. In general, inference happens in "real time;" for instance, object recognition in self driving cars, or speech-to-text and vice-versa transcriptions with an active user. Therefore, inference is latency-critical. On the other hand, training is mostly an offline workload. This opens up a new avenue to increase the efficiency of the computing hardware by employing a technique called *batched* stochastic gradient descent. Batching refers to an optimization where several inputs are collected together and then sent through the same computation pipeline simultaneously to enhance throughput. This technique is a natural fit for training since stochastic gradient descent already used sub-samples to update the weights (by performing backpropagation using the accumulated loss). Batching improves the efficiency by increasing the number of active units which would otherwise be inactive due to mapping constraints. Batching is usually not a good fit for inference, as the process of bundling together several camera frames increases latency and memory requirements, which can be critical resources.

Both training and inference are compute- and memory-intensive processes. Today, training is typically done on GPU clusters in datacenters, while inference is performed on both end-user devices (e.g., phones, cameras) and in datacenters on CPUs, GPUs, FPGAs, and custom accelerators. The scope of this book is the latter, i.e., custom accelerators for DNN inference.

1.1.2 DNN ARCHITECTURES AND LAYER TYPES

There can be myriad types of DNN topologies (also called architectures, but not to be confused with hardware architectures) depending on the composition of its layers. Figure 1.2 lists some popular DNN architectures in use today. The most fundamental DNN architecture is a multi-layer perceptron (MLP). A MLP consists of a series of fully connected (FC) layers, which are called as such as they connect every neuron's output as input to every neuron in the subsequent layer. Figure 1.1a shows a MLP. Other DNNs can be viewed as optimizations over MLPs. One of the most popular DNN architectures today is a convolutional neural network (CNN). CNNs use convolution (CONV2D) layers, which rely on connecting neurons to a limited continuous set of neurons in the next layer. This is known as a "receptive field," and is inspired by the visual cortex in animals. CNNs are built by a series of CONV2D layers followed by a classification layer at the end, which usually is a fully connected layer followed by softmax layer, and are thus popular for image processing tasks as such classification and segmentation. The receptive fields of different neurons partially overlap such that they cover the entire visual field of the image. Another popular class of DNNs is recurrent neural networks (RNN). RNNs introduce the notion of memory, which is often implemented by feeding neuron outputs back as inputs, and are valuable for speech processing tasks. RNNs are built using layers such as Long Short-Term

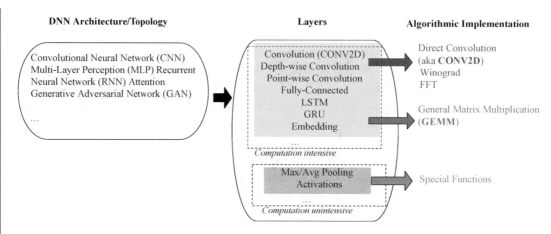

Figure 1.2: DNN layers and their implementations.

Memory (LSTM) and/or Gated Recurrent Units (GRU). Recently, DNNs with Attention [1] have attracted interest for natural language processing (NLP) and language translation tasks. Attention networks use embedding layers which encode words as dense matrices.

The DNN layers are the most computationally intensive parts of an entire DNN architecture, and are thus prime targets for acceleration. DNNs also employ other computations at the end of each layer, such as Max or Avg pooling, nonlinear activations, and so on. These can be viewed as separate computationally unintensive layers by themselves, or a sub-part of the computationally intensive layer itself.

1.1.3 POPULAR DNN MODELS

Over the last decade, researchers have proposed many DNN models every year. We list some of the state-of-the-art DNNs (at the time of writing this book) that have had a high-impact in Table 1.1 and describe them here. AlexNet [2] consists of five convolutional layers with grouped convolution and three fully-connected layers. Grouped convolution is partitioning channels into several groups to train the model in parallel in a multi-GPU environment. VGGNet [3] proposed deeper convolutional neural network (11–19 weight layers), demonstrating that deeper models can improve the accuracy. GoogleNet [4] presented the Inception operator that consists of branches of convolutions. In each branch, GoogleNet employs 1x1 convolution or pooling to reduce the computation overhead in the following convolutional layer with larger kernel size (3x3, 5x5, etc.). Resnet [5] introduced the skip connection, which adds a layer's activation value to the output activation of a later layer (identity operation). The skip connection was effective in addressing the vanishing gradient problem during training and increase training speed. Skip connections were later adopted in many other DNN models [6, 15]. MobileNetV2 [6] is a DNN model designed for mobile devices (or, edge devices). It provides good accuracy close to large

Table 1.1: Memory size* and computations in popular DNN models.

DNN Model	Application	Inputs (MB)	Weights (MB)	Outputs (MB)	Computations (GMACs)
AlexNet [2]	Image Classification	4.25	59.13	20.93	24.08
VGG 16 [3]	Image Classification	14.55	131.95	28.82	14.4
GoogleNet [4]	Image Classification	4.43	6.54	3.59	0.68
Resnet50 [5]	Image Classification	8.23	115.43	17.32	3.3
MobileNet v2 [6]	Image Classification	6.58	4.17	9.21	0.4
YoloV3 [7]	Image Classification	49.13	53.96	96.87	0.2
UNet [8]	Image Segmentation	2,033.74	26.93	3,573.93	2,428.9
DeepSpeech2 [9]	Speech Recognition	15.07	0.34	1.9	1.60
GNMT [10]	Language Translation	64.16	191.23	254.84	176.6
Transformer [11]	Language Translation	15	80.25	16.5	10.03
GPT2 [12]	Language Translation	8.75	19.27	23.75	19.26
NCF [13]	Recommendation System	3.88	1.08	4.96	0.6
DLRM[a] [14]	Recommendation System	0.005	2.225	0.004	0.002

[a] The memory sizes are shown for the MLP component of DLRM. DLRM also includes embedding tables that can be about 100 GB in size

* Assuming 8b inputs, 8b weights, and 16b outputs.

models (e.g., VGG16) while requiring significantly small memory size and computation. More recently, DNN models based on neural architecture search [16, 17] have been proposed. They optimize DNN models for both accuracy and efficiency (the number of computations). DNNs used for NLP have also gone through interesting innovations in the last few years. For speech processing workloads, networks like DeepSpeech2 [9] use a collection of recurrent and convolution layers. Recurrent layers are also used in language translation models such as GNMT [10]. However, the attention mechanism introduced by Transformer [11] is used in recent language translation networks like GPT2 [12] to gain significant improvements in accuracy on translation tasks. For recommendation workloads, which accounts for a majority of work among DNN applications in datacenters [18], state-of-the-art networks like DLRM [14] rely heavily on embedding layers.

Table 1.1 quantifies the computation and memory sizes across the DNN layers for some of the state-of-the-art DNN models discussed above. The computations are listed in terms of Giga Multiply-Accumulates (GMACs). We recommend taking a look at the MLPerf [19] benchmark suite to know and learn about more state-of-the-art DL models.

1.2   DNN ACCELERATORS

A key trend that is visible from Table 1.1 is that modern DNNs perform billions of computations and require tens to hundreds of MegaBytes of memory (for storing input, weight, and outputs). Moreover, applications using these DNNs on the cloud or edge often demand *real-time* inference latencies to take decisions from sensed/input data and display the results to users. In addition to tight latency constraints, DNNs running directly on edge devices (say smartphones running facial recognition) also come with tight energy constraints. CPUs are unable to meet real-time performance targets and energy constraints—as they typically house only tens of cores, and have power-hungry structures such as instruction schedulers and caches. Moreover, the end of Dennard Scaling and Moore's Law suggests that next-generation CPUs will maintain similar performance and energy budgets. GPUs, due to hundreds to thousands of ALUs on-chip, can run DNNs at orders of magnitude higher speeds than CPUs, at the cost of high energy consumption due to their inherent throughput-driven microarchitecture (state for hundreds of threads, and so on). Therefore, GPUs are mostly oriented at cloud-based training and inference, rather than consumer devices.

These constraints have led to a suite of specialized accelerators for DNN inference. These accelerators work on the well-known principle that a custom circuit for a specific problem can be orders of magnitude faster and more energy efficient than a general-purpose architecture. However, it is typically not economically viable to create custom circuits for every algorithm due to manufacturing cost and complexity. DL, on the other hand, is an extremely important application domain with high volume demands as AI/ML becomes pervasive. Therefore, custom silicon often turns out to be the economically superior solution. The combination of performance/energy and economic benefits has led to a rapid proliferation of DNN accelerators.

Figure 1.3 shows the template of DNN accelerator architectures assumed throughout the book, and the associated terms we will use. The architecture comprises of a *spatial array* of processing elements (PE) that house a multiply-accumulate (MAC) unit and a private local L1 buffer for inputs, weights, and outputs. The PEs are connected amongst each other and to a global shared L2 buffer via two networks: a distribution network to read inputs and weights, and a collection network to write the final outputs. Details of each of these components will be described through the rest of the book.

1.2.1   COMPUTATIONS WITHIN DNNS

Due to the nature of weighted sums, each neuron computation can be understood as a vector-vector dot-product. Thus, the computation of the collection of neurons within a layer—whatever the layer type might be—can usually be formulated as general matrix multiplication (GEMM) operations, as shown in Figure 1.2. For example, FC layers within MLPs/CNNs/RNNs, LSTM layers within RNNs, encoder/decoder within Attention networks [1], and so on are directly isomorphic to the dot-products in GEMM, as illustrated in Figure 1.4. GEMM operation also allows us to seamlessly employ optimizations like batching by altering the dimensions of the

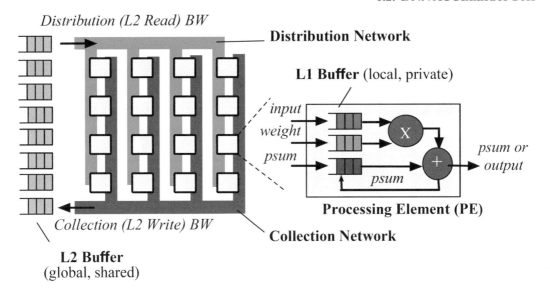

Figure 1.3: DNN accelerator template assumed throughout this book.

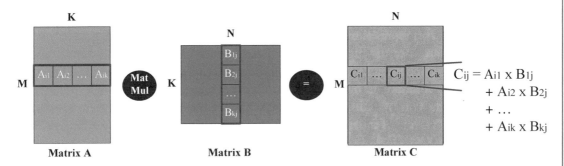

Figure 1.4: An illustration of general matrix multiplication (GEMM) operation. Multiplying Matrix A (MxK matrix) and B (KxN matrix) results in matrix C (MxN) matrix. For each element in matrix C, k partial sums (each element-wise multiplication result) need to be accumulated.

operand matrices. For the specific case of batching, extending the rows of the input operand matrix by concatenating it with the rows for multiple inputs would suffice. However, since our focus is on inference workloads, we will not go into such optimizations in detail.

Less obviously, convolution layers—which are one of the most common layers in CNNs—can also be represented as GEMMs, as we discuss in detail in Section 2.7. However, because convolutions have a restricted receptive field, which is often abstracted as a sliding window (similar to a stencil), convolutions are more naturally represented as a high-dimensional computation space, as Figure 1.5 shows. Roughly, this space corresponds to a matrix of planar convolutions

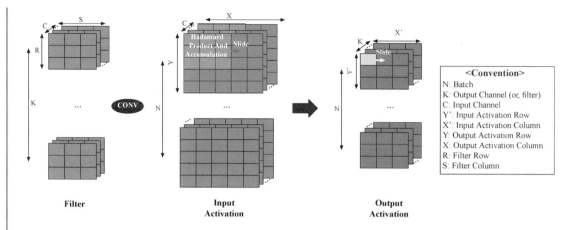

Figure 1.5: Visualization of CONV2D operation.

called *channels*. Input channels are reduced together (across the C dimension) into a single output channel (or plane). This is repeated with K different filters to form multiple output channels. This abstraction is based on the extension of the neuron abstraction we discussed earlier, resulting in a high-dimensional computation space based on three tensors[2] corresponding to input, weight, and output. This representation of the convolution layer is called as *direct convolution* or CONV2D[3] operation since the filter weight is planar in each channel. We illustrate a visualization of CONV2D operation in Figure 1.5 as a sliding kernel operation of each 3D filter on input activation. To compute an output, a Hadamard product (i.e., element-wise product) of all the elements in the corresponding 3D kernel needs to be computed, and then accumulated. CONV2D operation is one of the most common operations in recent DNNs (e.g., over 97% of computation is from CONV2D in Resnet50 [5]). Therefore, we will use CONV2D operation as the main example operation in this book, although the methods we present can also apply to GEMMs.

The weighted sum operation illustrated in Figure 1.1b is termed as a MAC operation. Depending on the DNN operator types, MAC operation can be between matrix and matrix (e.g., FC layer), vector and matrix, vector and vector, tensor and tensor (e.g., CONV2D), and so on.

1.2.2   CHALLENGE: DATA MOVEMENT

The fundamental roadblock to efficiency in modern computing systems, with DNN accelerators being no exception, is data movement. As Figure 1.6 shows, for a compute cost of 1x for two

[2]Data with multiple dimensions. A 1D tensor is a vector, and a 2D tensor is a matrix. This thinking can be extended to include scalars as 0D tensors.

[3]We use CONV2D interchangeably in this book to refer to either the convolution layer type, or its algorithmic implementation, as shown in Figure 1.2. The specific meaning can be interpreted from the context.

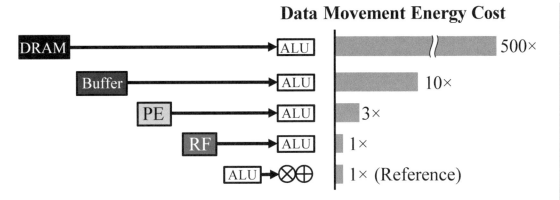

Figure 1.6: Data movement energy costs [20].

8-bit operands, the energy cost of fetching these operands to the compute units from remote on-chip SRAM is 10X, and from DRAM is 100–500x. This makes it crucial to (i) minimize the amount of data movement from DRAM to the accelerator, and also from the (ii) global on-chip SRAM to individual compute units, for energy efficiency. Reducing data movement is also crucial for performance, since bandwidth (both off-chip and on-chip) does not come for free. Unfortunately, as Table 1.1 shows, modern state-of-the-art DNNs with high accuracies have weight matrices in tens of MB to GB, which cannot completely fit on-chip, making data movement inevitable.

1.2.3 OPPORTUNITY: DATA REUSE

Although fetching the DNN model parameters and input activations from off-chip is inevitable, DNNs offer a unique opportunity in the form of data reuse that can be used to amortize the data fetching cost. For instance, in any matrix-matrix multiplication, each row of the first matrix gets reused by all columns of the second matrix, and each column of the second matrix gets reused by all rows of the first matrix. Moreover, unlike traditional workloads, the reuse opportunities within DNNs are very stylized as it comes from regular and well-defined operations like matrix multiplications and convolutions.

Dataflow and Data Orchestration. We define the maximum data reuse opportunity provided by the DNN as *algorithmic reuse*. For example, in a convolution operation, the same input activation gets reused across multiple input filters. The mechanism to leverage algorithmic reuse within a DNN accelerator is known as *dataflow*. Implementing a specific dataflow strategy in hardware requires clever *data orchestration*, which refers to the movement, retention, eviction, and synchronization of inputs, weights, and outputs, and is the key focus of this book. The actual data reuse we achieve is often termed as *arithmetic intensity*.

1.3   BOOK OVERVIEW

In the rest of the chapters, we tease out the various aspects of data orchestration. Chapter 2 provides background on data reuse and dataflows. Chapters 3 and 4 delve into buffer hierarchies and networks-on-chip, respectively, which form the key building blocks within DNN accelerators. Chapter 5 presents a systematic methodology and case studies on DNN accelerator design. Chapter 6 discusses design-space exploration of both the accelerator microarchitecture and mapping strategies. Chapter 7 introduces and discusses recent research in orchestrating sparse DNNs. Chapter 8 concludes this Synthesis Lecture with discussions on alternate platforms for DNN acceleration and some open research questions.

CHAPTER 2

Dataflow and Data Reuse

In this chapter, we first describe data reuse opportunities in common operations in DNNs. Since most of the realistic DNN operations involve millions to billions of computations, we cannot fit all of the computations within an accelerator, which typically has hundreds to thousands of compute units. Therefore, we need to slice the problem into smaller chunks (i.e., computation tiles) and run them in a certain order (i.e., tile scheduling). Within each computation tile, we also need to determine how to partition the computation across processing elements in an accelerator (i.e., spatial partitioning). We show how choices in tiling, scheduling, and partitioning affect the degree to which an accelerator can exploit the *algorithmic* reuse present in the original problem, and formalize these choices into the concepts of *dataflows* and *mappings*.

2.1 DATA REUSE OPPORTUNITIES

DNNs heavily rely on weighted sum operations that consist of MAC operations. These are abstracted as neurons as illustrated in Figure 2.1. Based on the layer type, the MAC operation can be between vectors, matrices, tensors, or some combination thereof. Among many layer types, we focus on convolution layers as they are heavily used in many DNNs, especially those targeted at image classification. Convolution layers involve tensor-tensor MAC operations, and provide ample data reuse opportunities along multiple dimensions of the tensors. These data reuse opportunities may be *temporal* (across time on the same compute unit, i.e., PE) or *spatial* (across compute units at the same time or across time). The amount of temporal and spatial data-reuse depends on the dataflow and mapping strategy, which we precisely define later in Section 2.3.

Figure 2.2a shows an example CONV2D operation and Figure 2.2b illustrates an example row-stationary [20] style mapping for this operation, and Figure 2.2c shows corresponding data and computation assignment on each PEs in the example mapping. We will discuss CONV2D in more detail in Section 2.6. Figure 2.2e shows all the data points accessed by each PE (y-axis) across time steps (x-axis). We can observe *replications of data points*. Such replication of accessed data points indicates data reuse opportunities revealed by a mapping. We highlight and label some of the temporal and spatial reuse opportunities in Figure 2.2e. These refer to the replicated accessed data points across time and space (PE), respectively, in the data usage timeline in Figure 2.2e. Spatial reuse opportunity can be further classified as purely spatial (multicasting in Figure 2.2d) or spatio-temporal (store-and-forward in Figure 2.2d). With proper hardware support, we can leverage such data reuse opportunities, and the resulting data reuse pattern of the example is as given in Figure 2.2d.

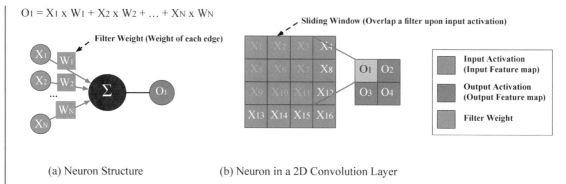

(a) Neuron Structure   (b) Neuron in a 2D Convolution Layer

Figure 2.1: (a) The fundamental neuron structure in DNNs, which performs weighted sum. Each edge has weight value multiplied with each input activation, or input feature map. Those multiplication results termed partial sums are accumulated to generate an output activation, or output feature map. (b) An example of neuron in a convolution layer. The region shaded in green over input activation is the filter weight applied over the overlapped input activation, which generates the first output activation highlighted in purple.

Data reuse opportunities within a DNN accelerator, such as the ones highlighted in Figure 2.2, are actually a subset of data reuse opportunities available within the neural network computation, which we term as algorithmic reuse. Algorithmic reuse from the target workload can be translated into specific data reuse opportunities via three architectural mechanisms within DNN accelerators—staging (i.e., store data in intermediate buffers and load it from the intermediate buffer, not global buffer, in the future), multicasting (i.e., simultaneously sending data to multiple PEs via shared wires; this reduces the number of buffer reads), and forwarding (i.e., sending data to an adjacent PE to use the data in the recipient PE, in the future). The purpose of staging is to keep data points in a local buffer to reuse them in the future, multicasting is to simultaneously send the same data points to multiple PEs to reuse them across PEs, and forwarding to send data points to adjacent PEs so that they can be reused in future iterations. Staging provides *temporal* reuse, while multicasting and forwarding provide *spatial* reuse. These can be implemented via buffer and interconnects, respectively. We categorize both multicasting and forwarding as spatial reuse because their data reuse is across space (PEs), and which of the two occurs depends on the implementation choices of the interconnects. We discuss these implementation choices later in Section 2.5.

To understand how the DNN computation intrinsically contains algorithmic reuse and how DNN accelerators can translate some of this algorithmic reuse into real data reuse, we provide a simple example, 1D convolution, and analyze algorithmic and actual data reuse of the 1D convolution in the following section.

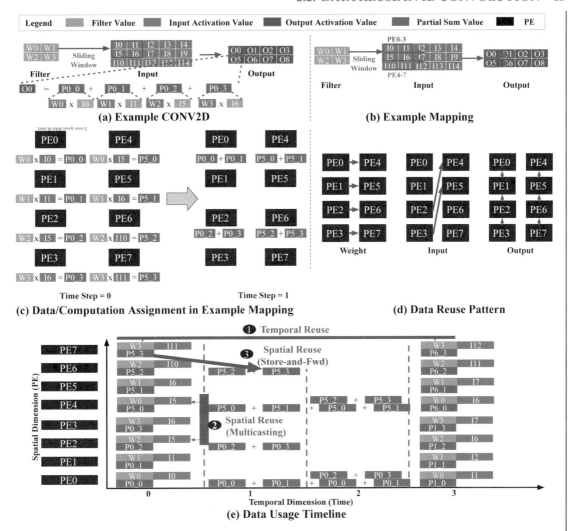

Figure 2.2: Three different data reuse styles in a CONV2D operation with no channels. (a) Shows the data labeling convention and the computation required for an output activation pixel. (b) Illustrates an example row-stationary data movement pattern [20]. (c) Shows temporal and spatial data reuse examples in an Eyeriss-style [20] accelerator with four PEs.

2.2 DATA REUSE IN 1D CONVOLUTION

For simplicity, we first analyze the simplest convolution operation shown in Figure 2.3a, which is called a 1D convolution or CONV1D operation. CONV1D does not include any height in input activations, height in filter weights, input/output channels (depth), input batches, or

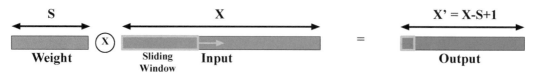

(a) 1D Convolution Operation in Output-stationary Style

for(x' = 0; x' < Bound(X'); x'++)   for(x = 0; x < Bound(X); x++)
  for(s = 0; s < Bound(S); s++)   for(s = 0; s < Bound(S); s++)
    O[x'] += W[s] * I[x'+s];   if(x-s < Bound(X) - Bound(S) && x-s > 0)
                                              O[x-s] += W[s] * I[x-s];

(b) Output-centric Representation   (c) Input-centric Representation

Figure 2.3: (a) A description of 1D convolution operation in sliding window operation over input activation and two loop nest versions of 1D convolution; (b) output-centric; and (c) input-centric. Note that both representations are interchangeable. This is an example of an *output-stationary* dataflow.

activation functions. 1D convolution provides three types of algorithmic reuse, which describe one-to-many relationship from a data point to computations.

- **Input Reuse:** Between adjacent sliding windows, there exist input data points (**halo**) in the overlap of sliding windows. Such input data points are required by computations in two (or more) of the adjacent sliding windows.

- **Filter Reuse:** The same set of filter weight values are required for computing all the output data points.

- **Output Reuse:** One output data point is the accumulation results of element-wise multiplication within a sliding window. We term each multiplication result as a unit partial sum. Since unit partial sums need to be accumulated to one final output value, the intermediate accumulation results can be reused during the accumulation process.

We discuss how this algorithmic reuse can be leveraged in a very simple accelerator with a single PE, first and then extend the discussion to the multi-PE case.

Single-PE case. In the 1D convolution example in Figure 2.3a, we first place weight filter values corresponding to the first set of input activations in the sliding window. We compute element-wise multiplication within the sliding window and accumulate all the multiplication results to produce one output activation pixel. After an output activation pixel is generated, we shift the sliding window by one pixel and perform the same computation to generate the next output activation pixel. We repeat this process until the sliding window reaches at the end of the

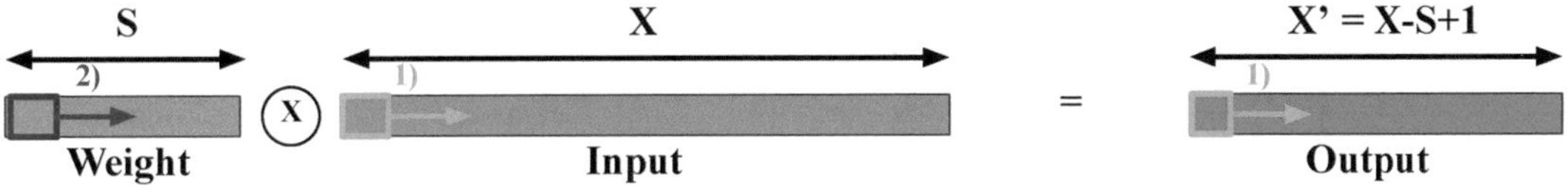

(a) 1D Convolution Operation in Weight-stationary Style

for(s = 0; s < Bound(S); s++)   for(s = 0; s < Bound(S); s++)
  for(x' = 0; x' < Bound(X'); x'++)   for(x = 0; x < Bound(X); x++)
    O[x'] += W[s] * I[x'+s];   if(x-s < Bound(X') - Bound(S) && x-s > 0)
                                             O[x-s] += W[s] * I[x-s];

(b) Output-centric Representation   (c) Input-centric Representation

Figure 2.4: An alternative style of computation for 1D convolution. Numbers over arrows refer the iteration order. The green boxes moves first; after the green box reaches the end of input feature map, the blue box moves one step. This is an example of a *weight-stationary* dataflow.

input activation generating all the output activation pixels. Such a computation schedule can be concisely represented using loop nests, as shown in Figure 2.3b and c. Both of the loop nests describe the same computation and schedule but (b) employs output index while (c) employs input index. This is because input and output activation indices are mutually dependent so we need to select one of them to describe the other. We term those two representation styles as input- and output-centric loop nests, respectively.

Note that the loop nest in Figure 2.3 is just one of the possible styles to compute 1D convolution. We call this an *output-stationary* style loop nest, because the output activation index is updated in the slowest manner [20]. Alternately, if we interchange the loop order (loop interchange) in Figure 2.3b, we obtain a new version of loop nest, as described in Figure 2.4b, that generates the same output activation values. In the new loop nest, the weight index changes in the slowest manner because the weight index loop (loop S) is placed at the upper most level. Therefore, this new loop nest is called a *weight-stationary* style. Such combinations of loop transformations are termed as **dataflows** [20]. We describe dataflows formally in Section 2.3.

Although both of the loop nest styles compute the same output activation values, they imply different buffer size requirements and buffer access counts which imply dramatically different energy consumption based on the actual dimension size (S and X in the example). Figure 2.5 illustrates this point. For a simple single PE accelerator with a local PE buffer connected to external DRAM running the 1D convolution operation, Figure 2.5c summarizes algorithmic reuse, maximum data reuse of each data class, minimum buffer sizes to achieve the maximum data reuse, and minimum DRAM access counts when the maximum data reuse is achieved. The two different dataflow styles (output and weight-stationary) minimizes DRAM accesses of one of the data classes (output and weight, respectively) and corresponding PE buffer sizes but have

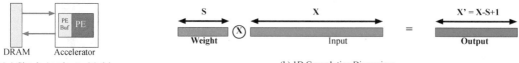

(a) A Simple Accelerator Model   (b) 1D Convolution Dimensions

Data Reuse Strategy	Algorithmic Minimum DRAM Accesses			Maximum Operand Uses			Minimum PE Buffer Size for Zero Re-fetch			Minimum DRAM Accesses		
	Weight	Input	Output	Weight	Input	Output	Weight	Input	Output	Weight	Input	Output
Output Stationary	S	X	X'	SX'	SX'	SX'	S	S	1	SX'	SX'	X'
Weight Stationary							1	X'	X'	S	SX'	SX'

(c) Cost Summary

Figure 2.5: The impact of dataflow on memory (PE buffer/DRAM) access and size costs.

(a) 1D Convolution Operation in Output-stationary Style

```
for(x'2 = 0; x'2 < Bound(X'2); x'2++)
  parallel_for(x'1 = 0; x'1 < 2; x'1++)
   for(x'0 = 0; x'0 < 1; x'0++)
    for(s = 0; s < Bound(S); s++)
     x' = x'2*1*2 + x'1*1 + x'0
     O[x'] += W[s] * I[x'+s]
```

	DRAM	Accelerator
		PE0 PE1
Time 0	O[0]	O[1]
Time 1	O[2]	O[3]
...	...	...

(b) Output-stationary Loopnest on Two PEs   (c) Output Mapping on Two PEs

Figure 2.6: An example of dataflow on multiple PEs. (a) Shows the 1D convolution sizes with the first data mapping on PE0 and PE1. (b) Shows a loop-nest representation of the example output-stationary dataflow. We assume the bound of loop x'0 is 1, for simplicity. (c) Describes the parallelism over output activations in the example dataflow.

high DRAM accesses and buffer sizes for the other data classes. For example, output stationary matches the algorithmic minimum DRAM access of outputs but does not provide any benefits for weight and input tensors. As the minimum DRAM access numbers show, the overall efficiency of the two dataflow styles will depend on the problem dimension (S and X').

Multi-PE case. In previous 1D convolution examples, we analyzed data reuse over one PE over time, or temporal data reuse. When an accelerator has multiple PEs to exploit parallelism as most of spatial accelerators do, the accelerator can have more complex data reuse

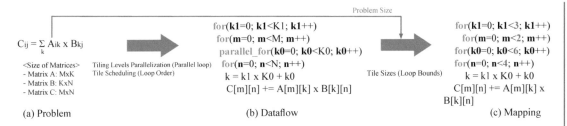

(a) Problem   (b) Dataflow   (c) Mapping

Figure 2.7: Loop nest representation of an example matrix multiplication problem, dataflow, and mapping.

since the data reuse can occur spatially, or across PEs at the same time, via multicasts. For example, Figure 2.6 shows an output-stationary style dataflow over an accelerator with two PEs in Figure 2.6c. Unlike single PE cases in previous examples, this example shows spatial reuse opportunities in input activation implied by the overlapped region (halo) between data mappings on PE0 and PE1 in Figure 2.6a. The overlapped data, or halo, imply that they can be multicasted to each PE to reduce DRAM accesses.

2.3 DATAFLOWS AND MAPPINGS

In the previous section, we saw how even a simple CONV1D operation can exhibit multiple forms of temporal and spatial reuse depending on the manner in which the computation was tiled, scheduled and partitioned across the hardware resources.

We refer to a specific codification of these choices as a *dataflow*. One such codification we use is a loop nest representation similar in form to the loop nests we have used in prior sections. The information encoded in this representation is the loop order, choice of loops to parallelize, and the number of loop tiling levels. However, the loop bounds themselves are left as symbolic names. This means that a dataflow prescribes the tiling, scheduling, and partitioning strategy, but not the specific *numbers* of tiles, partitions or loop iterations. Thus, a dataflow expresses an execution strategy for a general problem shape (such as a CONV2D layer) but without explicit workload size information (i.e., size of each data dimensions in input/output tensors of the workload).

A *mapping* is a specific *instance* of a dataflow with numeric (instead of symbolic) loop bounds. Thus, a mapping precisely describes the execution of a specific workload instance, such as the CONV2D layer *VGG16_3_2*.

Figure 2.7 shows loop nests that represent an example matrix multiplication problem, an example dataflow, and a mapping based on the problem and dataflow. Figure 2.8 illustrates the overall process of mapping a CONV2D operation onto an example accelerator.

It stands to reason that any reasonable hardware accelerator must have sufficient configurability to support mappings for a variety of workload instances—even though all those mappings

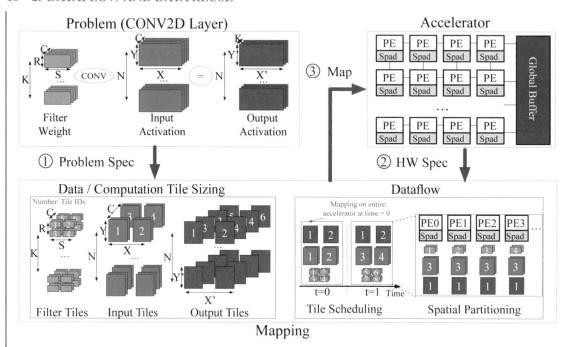

Figure 2.8: A high-level overview of the process of mapping a problem onto an accelerator.

may simply be instances of the same dataflow. However, many *flexible* hardware accelerators in fact support multiple dataflows. This is especially true for accelerators with multiple levels in their storage hierarchy. Sections of the dataflow corresponding to the innermost storage levels are often *baked* into the hardware for efficiency, but the outer levels of the hierarchy may consist of programmable state machines, allowing the hardware to run more than one dataflow.

2.4   DEEP DIVE INTO DATAFLOWS AND MAPPINGS

In Figure 2.9 and Figure 2.10, we build six example mappings upon the simple 1D convolution discussed in Section 2.2 to demonstrate how simple changes to a mapping expose various forms of reuse—both spatial and temporal.

The impact of parallelization. Parallel for loops, denoted as pfor in Figure 2.9 and Figure 2.10 specify the spatial data tile distribution across PEs. For example, in mapping A, pfor(x0=0; x0<3; x0++) (where X' refers to the first dimension of output vector or output width), spatially distributes indices of the X' dimension with a tile size of one across PEs. The number of iterations in a parallel for loop needs to exactly match the total number of PEs. When the loop nest has many parallel for loops, the product of the each loop bound needs to match the total number of PEs.

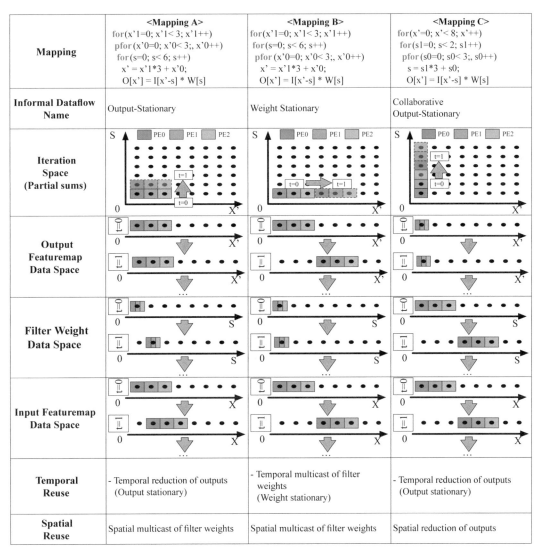

Figure 2.9: The first three of six mapping examples for CONV1D operations. The remaining three are shown in Figure 2.10. Mappings A, B, C, and D show the impact of loop order. Mapping E shows the impact of different tile size. Mapping F shows the impact of multi-level parallelism with tiling. **pfor** indicates a parallelized for loop (parallel_for) in this figure.

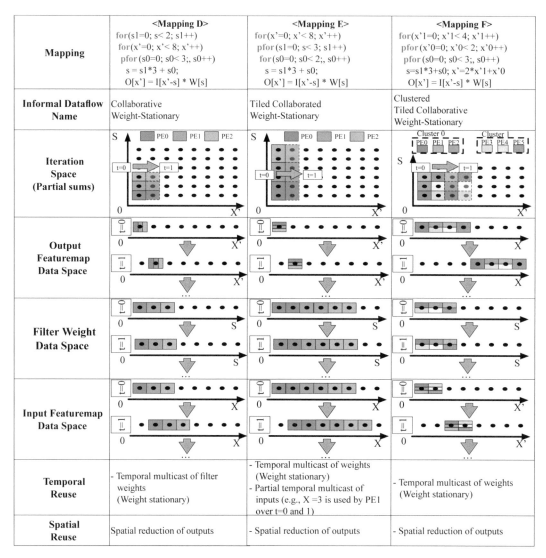

Figure 2.10: The last three of six mapping examples for CONV1D operations, continued from Figure 2.9.

Similarly, a temporal for loop denoted as plain "for" in loop nests specifies the distribution of a dimension across time steps in each target PE. That is, temporal for maps the same set of indices for a dimension across PEs. For example, `for(s=0; s<6; s++)` of mapping A in Figure 2.9 (where `S` refers to the first dimension, or width, of filter weight data class), distributes indices of the `S` dimension with a chunk size of one across time steps. This temporal distribution can be viewed from the data space of the filter weight data class in the same column of mapping A in Figure 2.9. Since all PEs get same data indices corresponding to a temporally mapped dimension, this may create opportunities for *spatial reuse*, i.e., spatially reusing the same data values across PEs in a time step.

The impact of loop order. The order of parallel and temporal for loops in a loop nest dictates the order of data movement, which also changes the data mapping to PEs across time. A change in the order can result in an entirely different stationary behavior. For example, the sequence of loops in mapping A in Figure 2.9 (i.e., parallel for on `X'` followed by the temporal for on `S`) indicates that all data indices of S should be explored before working on the next chunk of `X'` indices. This order results in temporal reuse data corresponding to `X'` indices (i.e., partial sums, for all indices of `S`) leading to an output stationary dataflow. This behavior can be observed from the iteration space for the mapping A in Figure 2.9.

If the loops over x'0 and s are interchanged as shown in mapping B in Figure 2.9, the dataflow of the resulting mapping now keeps weights stationary because PEs can temporally reuse data corresponding to `S` indices, (not `X'` indices anymore), (i.e., weight values, for all indices of `X'`) before moving to the next set of `S` indices. Similarly, mappings C and D in Figure 2.9 and Figure 2.10 show the spatial distribution on `S` instead `X'`, and also the impact of data movement order on temporal reuse leads to different stationary dataflows.

The impact of tiling. In all of the mapping styles from A–D in Figure 2.9 and Figure 2.10, the tile sizes are all one—resulting in either no temporal reuse (e.g., partial output sums in case of mapping B) or full temporal reuse (e.g., input feature map in mapping C). Increasing the tile size of the parallel and temporal for loops can help in capturing partial temporal reuse opportunities, or convolutional reuse in input feature map of convolution layers. For example, the parallel for on the `S` dimension in mapping E in Figure 2.10 enables partial temporal reuse of input feature map data across time steps.

Also, tiling can be used to divide iterations into smaller units as shown in mapping D where the loop over dimension `S` is split into two loops and placed across loop over dimension `X'`, which indicates that we do not have to fully explore indices of `S` dimension before we change mapped indices on `X'` dimension. This not only removes fundamental restriction of loop nests that require all the inner loops complete before updating the loop variable of a loop but also reduces the data tile size to fit into on-chip global and PE buffers.

Exploiting multi-dimensional spatial distributions. To exploit multi-dimensional parallelism, a loop nest can employ multiple parallel for loops like mapping F in Figure 2.10. Each parallel for loop indicates a parallelization dimension, which eventually implies dimensionality in a PE array. For example, mapping F in Figure 2.10 has two parallel for loops and implies two clusters of PEs with three PEs in each, as shown in iteration space row of the example. However, if two parallel for loops are specified on an overlapped dimension (e.g., input feature map column and filter column), it does not imply additional dimension in a PE array. A loop nest needs to ensure the product of loop bounds of each parallel for matches the number of PEs. If the numbers do not match, the mapping results in either unavailable to map on a PE array (when the product of loop bounds, or *the degree of parallelization*, is larger than the number of PEs) or PE underutilization (the degree of parallelization is smaller than the number of PEs). This is because the degree of parallelization is the total amount of parallelism implied by the data orchestration and the number of PEs is actually available parallelism in hardware. If the numbers do not match, it implies that the mapping does not properly fit onto the target hardware.

As observed in the examples of dataflows and mappings of a 1D convolution introduced in Figure 2.4, loop order and resulting stationary behavior based on dataflow play a key role to materialize algorithmic reuse into an actual reuse in computing schedule. However, to realize such algorithmic reuse in hardware, appropriate hardware support is required. For example, the network-on-chip (NoC) between the DRAM and the PE array in Figure 2.6c needs to support multicasts to exploit spatial data reuse opportunities implied by the dataflow in hardware. Without the support for multicasts from NoC, the spatial reuse opportunity is forfeited and translated into independent unicasts, which do not reduce DRAM accesses at all in the example architecture of Figure 2.6c. To discuss the hardware support for each data reuse type, in the next section, we classify the data reuse in more fine-grained manner and discuss hardware choices to actually exploit the data reuse in the accelerator.

2.5   HARNESSING DATA REUSE VIA HARDWARE SUPPORT

As we discussed above, various data reuse opportunities appear based on the dataflow. Table 2.1 summarizes how such opportunities appear in the relationship of spatially mapped dimension and inner-most temporally mapped dimension. For example, if output channels (K) are spatially mapped (i.e., parallelized), input and output activations do not change over space, since input and output activations are not coupled with the output channel dimension (K). That is, all the PEs receive the same input feature map, which implies a full spatial reuse opportunity (broadcast). In the same example, when the inner-most temporally mapped dimension is the input channels (C), the input channel changes every iteration, which provides temporal reduction opportunities of outputs.

Although a dataflow provides temporal or spatial data reuse opportunities, appropriate hardware support is required to actually exploit those reuse opportunities. Table 2.2 summarizes four reuse categories and corresponding hardware implementation to support them. As the table shows, data reuse can be either spatial or temporal. Based on the data class, the communication type can be either multicast (input tensors) or reduction (output tensors). Multicast is a com-

Table 2.1: Reuse opportunities based on spatially mapped dimensions in combination with innermost temporally-mapped dimensions. Filters (F), Inputs (I), and Outputs (O) are considered separately. For brevity, X/Y should be interpreted as X'/Y' as appropriate.

| Spatial | | | | | | | Temporal | | | | | | |
Mapped Dimension	Coupling F	Coupling I	Coupling O	Reuse F	Reuse I	Reuse O	Innermost Mapped Dimension	Coupling F	Coupling I	Coupling O	Reuse F	Reuse I	Reuse O
K	✓		✓		Multicast		C	✓	✓				Reduction
							R/S	✓		✓		Multicast	
							X/Y		✓	✓	Multicast		
C	✓	✓				Reduction	K	✓		✓		Multicast	
							R/S	✓		✓		Multicast	
							X/Y		✓	✓	Multicast		
R/S	✓		✓		Multicast		K	✓		✓		Multicast	
							C	✓	✓				Reduction
							X/Y		✓	✓	Multicast		
X/Y		✓	✓	Multicast			K	✓		✓		Multicast	
							C	✓	✓				Reduction
							R/S	✓		✓		Multicast	

munication type that delivers the same data to multiple targets over space (different PEs at the same time) or time (the same PE in different time). Therefore, multicast is one to many communication type, which requires either a fan-out NoC structure such as bus or tree, or a "stationary" buffer to hold the data and deliver it to the future. In contrast, reduction is many to one communication type, which applies to partial sums to generate final outputs. Reduction also can be either temporal or spatial. Temporal reduction can be supported by a read-modify-write buffer. Example hardware to support spatial reduction is reduction tree or reduce-and-forward chains. We discuss detailed implementations (and hardware costs) for the temporal (i.e., buffer-based) and spatial (i.e., interconnect-based) reuse structures in Chapter 3 and Chapter 4, respectively.

In summary, different dataflow styles expose different forms of reuse: spatial and temporal, both for multicasts and reductions, which in turn can have multiple hardware implementations. Reasoning about dataflow in this structured manner exposes new insights, and potential microarchitectural solutions. To provide deeper understandings of the analysis, we provided detailed examples of 1D convolution dataflows that covers all the possible types without multi-level tiling. Next, we move on to the full convolution operation.

Table 2.2: Hardware implementation choices for supporting spatial and temporal reuse. Note—by *temporal multicast*, we refer to *stationary* buffers from which the same data is read over time.

Reuse Type	Communication Type	HW Implementation Choice	
Spatial	Multicast	Fanout (e.g., Bus, Tree)	Store-and-Fwd (e.g., Systolic Array)
	Reduction	Fanin (e.g., Reduction Tree)	Reduce-and-Fwd (e.g., Systolic Array)
Temporal	Multicast	Multiple reads from a buffer	
	Reduction	Multiple read-modify-write to a buffer	

2.6   DATAFLOWS AND DATA REUSE IN CONV2D

CONV2D operation is one of the most common operations in CNN that involves seven data dimensions across three tensors: input/output activation and weight tensors. Unlike CONV1D, CONV2D operation involves seven data dimensions, which makes the CONV2D operation a high-dimensional operation. Therefore, we first introduce the CONV2D operation and then discuss the data reuse opportunities and dataflow choices.

2.6.1   CONV2D OPERATION

Recall that CONV1D computes an output vector (output activation) using two input vectors (input activation and filter) as operands. The CONV2D operation computes an output **tensor** (output activation) using two input **tensors** (input activation and filter) as operands. To understand the high-dimensional CONV2D operation, in Figure 2.11, we construct the full version of CONV2D operation from the simplest form, which we name as plain CONV2D. Figure 2.11a lists the labeling conventions for each data dimension.

Plain CONV2D. Assuming single-layer CNN, input and output activation can be viewed as input and output images of an image processing program (e.g., blurring), and filter can be viewed as an image processing mask (or, kernel). Assume that the images are grayscale images, then the simplest CONV2D operation upon the grayscale images can be visualized as the plain

Data Dimension	Batch	Output Channel	Input Channel	Filter Row	Filter Column	Input Row	Input Column	Output Row	Output Column
Notation	N	K	C	R	S	Y	X	Y'	X'

* Uppercase: Size, Lowercase: index

(a) Data Dimension Labeling Conventions

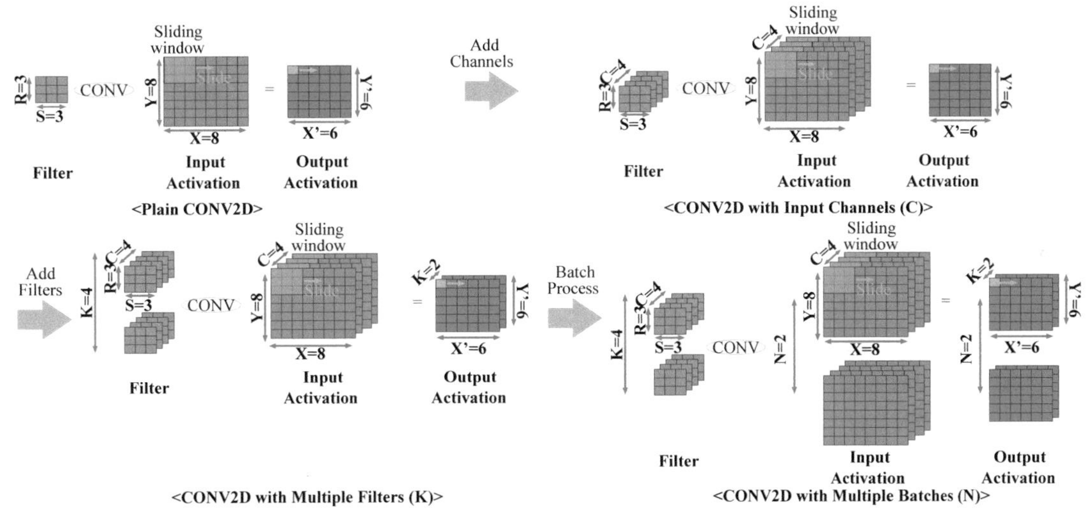

(b) Constructing the Full CONV2D Operation from Plain CONV2D

```
for(n=0; n<2; n++)
 for(k=0; k<2; k++)
  for(c=0; c<4; c++)
   for(y=0; y<8; y++)
    for(x=0; x<8; x++)
     for(r=0; r<3; r++)
      for(s=0; s<3; s++)
       if(y-r >= 0 && x-s >= 0)
        O[k][y-r][x-s] += W[k][c][r][s] * I[c][y][x];
```

(c) Input-Centric Loop Nest

```
for(n=0; n<2; n++)
 for(k=0; k<2; k++)
  for(c=0; c<4; c++)
   for(y'=0; y'<6; y'++)
    for(x'=0; x'<6; x'++)
     for(r=0; r<3; r++)
      for(s=0; s<3; s++)
       O[k][y'][x'] += W[k][c][r][s] * I[c][y'+r][x'+s];
```

(d) Output-Centric Loop Nest

Data class \ Data Dim.	Batch (N)	Output Ch. (K)	Input Ch. (C)	Filter Row (R)	Filter Col. (S)	Input Row (Y)	Input Col. (X)
Output Activation	✓	✓		✓	✓	✓	✓
Input Activation	✓		✓			✓	✓
Filter Weights		✓	✓	✓	✓		

* Output row(Y') = Y-R+1, Output column(X') = X-S+1

(e) Data class and Coupled Dimensions in Input-centric Loop Nest

Figure 2.11: An example of a convolutional layer with its dimensions and indexing are shown in (a), and a visualization of the convolution shown in (b). An input-centric and output-centric view of loop nests corresponding to the convolution is shown in (c) and (d), respectively. A summary of the coupling among dimensions and data classes (tensors) are shown in (e), where a table entry with a check mark indicates that the dimension in the column is coupled with the data class in the row.

CONV2D in Figure 2.11b. Note that the sliding window is now 2D unlike the CONV1D operation; thus, the operation is termed as CONV2D. The 2D sliding window in CONV2D needs to sweep the entire surface of the input activation. That is, the sliding window moves along both of the input row (Y) and column (X) dimensions, performing the same element-wise multiplication and accumulation as we computed in CONV1D operation.

CONV2D with input channels. Just like images typically have red-green-blue (RGB) channels, input activation can also have channels. When input activation has multiple channels, filters need to have the same number of channels as that of input activation, so that all the input activation channels have corresponding filter values. That is, the sliding window is now three-dimensional, as shown in CONV2D with input channels in Figure 2.11b. Therefore, the element-wise multiplication results are accumulated not only in filter row and column dimensions but also in the input channel dimensions, visualized as a depth dimension in Figure 2.11b. The width and height of the sliding window are referred as filter row (R) and height (S) dimensions. The width and height of the input activation where the sliding window sweeps are referred as input row (Y) and column (X) dimensions.

CONV2D with input channels and multiple filters. A 3D filter with input channels extracts one feature from the input activation. When we need to extract multiple features, we apply multiple 3D filters. Since we have multiple filters, we obtain multiple and independent set of output activation values, constructing the output channel (K) dimensions in output activations, as shown in CONV2D with multiple filters in Figure 2.11b. We can view this as swapping the filters in the previous version (CONV2D with input channels), generating multiple set of output activation values and arranging them as the depth (i.e., channel) of the output activation.

CONV2D with channels and multiple batches. We often process multiple input and output activations during inference. Instead of processing them one-by-one, we can construct a batch (N) of the input activations and run all at once. Such scenario is illustrated in CONV2D with multiple batches in Figure 2.11b. This can be viewed as we are running multiple instances of the previous version, CONV2D with input channels and multiple filters, and arranging the inputs and results as the sequence of 3D input and output activation tensors, like illustrated as N dimension in the last step of Figure 2.11b. This version is the full CONV2D operation that can be easily found in recent CNNs such as Resnet [5].

Loop-nest representation of full CONV2D. Using a subset of the data dimensions, we can represent the computation of CONV2D in a loop nest. Figure 2.11c and d shows two possible loop nest representation of the example CONV2D operation in Figure 2.11b. Two versions exist since the row and column indices of input/output activation tensors are mutually deducible. For example, the input row index (y) corresponds to given output row index (y') and filter row index (r) can be computed $y = y' + r$, as shown in Figure 2.11d (e.g., if we process the first output row ($y' = 1$) and the second filter row ($r = 2$), we need to access inputs in the third row ($y =' y + r = 1 + 2 = 3$)). Based on the choice of input and output row/column indices to use the in loop nest representation, we can obtain two versions of loop nests. Like we did in the CONV1D examples, we name those two versions input- and output-centric loop nests, respectively.

2.6.2   DATA DIMENSION COUPLING AND DATA REUSE OPPORTUNITIES

When a data dimension exists in a tensor, then the data dimension is *coupled* with the tensor. Figure 2.11e shows how seven data dimensions in the CONV2D operation are coupled with three tensors.

Data dimensions other than activation row and column do not have such coupling but they can be coupled with multiple tensors or only one tensor. For example, the input channel index c appears in both filter and input activation, and the output channel k appears in both filter and output activation. We call these dimensions *coupled* to these indices, as the position in the data space changes when the index is modified. That is, when a tensor A is independent of a dimension α, the mapping of the tensor is stationary when α is updated in a loop nest. This implies the possibility of temporal data reuse via proper loop order. From another perspective of spatial mapping, when we parallelize the loop on the dimension α, the tensor A is stationary across **space** (i.e., PEs). This implies the possibility of spatial data reuse via multicasts.

Likewise, we can exploit the dimension coupling via loop order and parallel loops to expose the data reuse opportunities. In addition, we can also control the amount of data reuse via loop tiling. Since the CONV1D operation discussed in Figure 2.9 does not show complex dimension coupling relationship, we provide a realistic dataflow example in Figure 2.12 and discuss the data reuse in the example.

2.6.3   DATA REUSE IN A CONV2D EXAMPLE

Figure 2.12a and b describe an example mapping based on the row-stationary dataflow introduced in Eyeriss [20] on an example convolution shown in Figure 2.11b. Figure 2.12c describes actual mapping of tensors on an accelerator with six PEs. The six PEs are clustered into two groups of three PEs each because the loop nest has multiple parallel for loops. The clustering dimensionality follows the loop bounds of each parallel for. Note that we refer each computation tile iteration upon the PE array in the accelerator as a time step, which corresponds to the iteration of input activation tile (loop x1), as we denote in Figure 2.11a. Therefore, we refer to loop x1 as the unit loop of this example.

The mapping places the loop over activation row tile $y1$ above the unit loop (loop x1), which makes the activation row stationary (i.e., row-stationary) across time steps. We deep dive into the data reuse pattern for each of three tensors: input activation, filter, and output activation tensors.

Input Activation. Each of the PE clusters receives three input activation rows with stride one. The input activation column dimension is fully covered within the inner loops under the unit loop. Also, loops on other coupled dimensions (input channel c and batch n) with input

```
for(n=0; n<2; n++)
 for(c1=0; c1<2; c1++)
  for(c0=0; c0<3; c0++)
   for(k1=0; k1<2; k1++)
    for(k0=0; k0<2; k0++)
     parallel_for (y1=0; y1<2; y1++)
      for (x1=0; x1<5; x1++)    ←——— Time step in
                                       this example
       parallel_for (y0=0; y0<3; y0++)
        for(x0=0; x0<3; x0++)
         k=2*k1+k0; c = 3*c1+c0; r = y0; s = x0;
         y=stride*y1 + y0; x=stride*x1 + x0;
         P[n][k][c][y-r][x-s][r][s] = W[k][c][r][s] * I[n][c][y][x];
         O[n][k][c][y-r][x-s] += P[n][k][c][y-r][x-s ][r][s];
```

(a) Loop nest representation

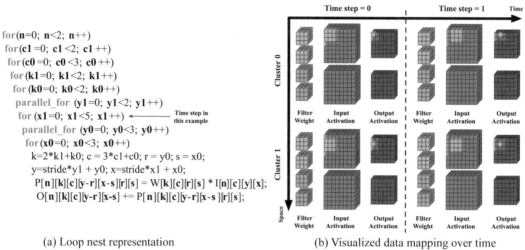

(b) Visualized data mapping over time

(c) Data mapping and reuse of each tensor

Figure 2.12: Detailed mapping description of an accelerator with six PEs running a mapping based on row-stationary style [20] dataflow. The colors in (b) represent each tensor and a computation tile, and that in (c) represent replicated data (i.e., data reuse opportunities) from the mapping. We refer to computation tile iterations on the PE array in the example accelerator as timesteps in this example to emphasize the temporal scheduling of computation tile mapping. We mark the loop nest corresponding to the computation tile iteration (or, timesteps in this example) in (a).

activation are placed in the upper positions of the unit loop. Therefore, those three dimensions $(n, c, \text{and } x)$ are stationary. However, because we have stride of one in activation row dimension and we parallelize the activation row across PEs within a PE cluster in order, the replicated input activation is skewed as shown in the first column of Figure 2.12c. Therefore, we can observe a spatial data reuse opportunity on input activations in a diagonal direction of the PE array.

Filter. Each of the PE clusters has full filter row (r) and column (s), and all the loops on input channel (c) and output channel $(k$ are placed in the outer loops of the unit loop. Because filter is coupled with those four dimensions (k, c, r, s), the filter is stationary across PE clusters. Within a PE cluster, since we parallelize filter rows along $y0)$, we distribute three filter rows across three PEs in the PE cluster. This results in horizontal replication of filter values as shown in filter weight column in Figure 2.12, which implies a spatial data reuse opportunity.

Output Activation. Because the batch (n) and output channel (k) dimensions are placed in upper loops of the unit loop, and activation column is fully mapped in the inner loops of the unit loop, those three dimensions $(n, k, \text{and } x')$ are stationary across PE clusters. However, the output activation row is not stationary across PE clusters. Each PE cluster receives three consecutive input activation rows, and this can be translated as one output activation row because the filter row size is three. Because of the parallel for on activation (loop $y1$), each cluster receives different output activation row with stride one. That is, each cluster processes different output rows. Within each PE cluster, three PEs collaborate to generate partial sums and spatially accumulate them to produce output activation values. For the accumulation, partial sums flow linearly within each PE cluster, as shown in the third column of Figure 2.12c.

2.7   CONVOLUTION AS MATRIX MULTIPLICATION

On devices that have highly optimized architecture and compiler support for GEMM or matrix multiplication operations (e.g., GPUs and Google TPU), convolution operations are converted into matrix multiplication operations to exploit the underlying infrastructure [21]. To convert a convolution operation into a matrix multiplication, we flatten input activation and filter and reorganize them into two matrices. For example, the *im2col* style transformation [21, 22] flattens entire filter and input activation within sliding windows (i.e., receptive fields), as shown in the example in Figure 2.13.

We use the following convention for the GEMM operation: the input matrix has dimensions $M \times K$, the weight matrix has dimensions $K \times N$, and the output matrix has dimensions $M \times N$.[1] The transformation performs the following steps to convert a CONV2D into a corresponding GEMM operation.

1. Flatten receptive fields on input activation into column vectors (i.e., image to column vectors or im2col).

---

[1]The dimensions K and N in the GEMM notation are different from K and N in the CONV2D notation. Figure 2.13 shows the relationship.

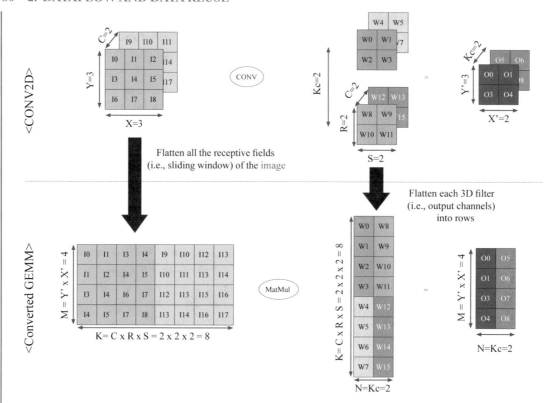

Figure 2.13: im2col transformation of a CONV2D operation into a corresponding GEMM operation. When batches exist, receptive fields (i.e., input activation values in possible sliding window positions across input channels) of batches are concatenated to the N dimension (column) on the second matrix. Corresponding outputs are concatenated to the N dimension of the output matrix.

2. Concatenate the column vectors and construct the first matrix ($M - K$) of the converted GEMM as shown in Figure 2.13.

3. Flatten each filter (i.e., three-dimensional filter data with input channel (C), filter height (R), and row (S)) into a row vector.

4. Concatenate the row vectors of each filter to construct a filter matrix, and transpose it, as shown in the second matrix ($K - N$) of converted GEMM in Figure 2.13.

When a convolution is converted into matrix multiplication, data reuse in flattened tensor becomes more explicit since we perform a set of dot products using the flattened vector (e.g., replicated input activation values). However, the reuse between the halos (i.e., sliding window) gets lost.

2.8   SUMMARY

In this chapter, we discussed (1) how we define dataflow and mapping and (2) how mapping and dataflow affect data reuse opportunities using CONV1D and CONV2D examples. We also highlighted that appropriate hardware support is necessary to exploit data reuse opportunities in accelerators. In the following chapters, we discuss the details of key hardware components to support various dataflows: buffers and on-chip networks.

CHAPTER 3

Buffer Hierarchies

A key component of data orchestration is the staging *buffer hierarchy*. Although memory hierarchies have been well studied in general-purpose computers, domain-specific accelerators have constraints and goals that differ in key ways. It is important to understand in detail how these cause accelerator architects to make different hardware choices. In this chapter, we present a framework for understanding key options, and explore tradeoffs between design effort and cross-project reuse.

3.1 MOTIVATION

Customized accelerators enable architects to leverage static workload and domain knowledge to improve efficiency. Recent accelerators have made significant strides in custom datapaths, novel algorithm tweaks, and data representations [20, 23–29]. However, accelerators do nothing to address one of the most fundamental problems in computer architecture: the "off-chip memory" wall. Dally reports [30] that the relative difference in energy and latency between arithmetic, on-chip data movement, and off-chip accesses can be up to 800× for 20 nm process. Details are reproduced in Table 3.1.

Furthermore, the custom datapaths and algorithms employed by accelerators can actually exacerbate this problem—improvements in data processing efficiency can increase the rate at which new data is required from off-chip. (To counteract this, accelerators spend significant area for on-chip buffer hierarchies, as shown in Table 3.2.)

A key difference from general-purpose processors is that accelerator architects can leverage design-time knowledge of the workload and domain. This can allow the buffer hierarchy to be tailored exactly to the needs of the accelerator, sizing buffers, and distributing bandwidth for optimized performance/efficiency tradeoffs. However, this comes at the cost of increased design effort, which can harm time-to-market and lead to a wasteful duplication of engineering effort across designs. Less obviously, the lack of a consistent abstraction for data movement, staging, and synchronization also means a dearth of toolflows exist to auto-orchestrate datasets and generate control Finite State Machines (FSMs) for data movement. Therefore, there is a particular need for the accelerator architecture community to develop cross-design buffering abstractions and practices, especially for fast-moving application domains like machine learning, lest accelerators find themselves obsolete before reaching the market.

Table 3.1: Energy cost estimates for 20 nm from Dally [30]

64b math op	20 pJ	1 ×
8KB SRAM: 256b access	50 pJ	2.5 ×
256b bus, 3 mm	26 pJ	1.3 ×
256b bus, 9 mm	256 pJ	12.8 ×
256b bus, 40 mm	1,000 pJ	50 ×
Efficient o-chip link	500 pJ	25 ×
DRAM Read/Write	16,000 pJ	800 ×

Table 3.2: Percentage of area dedicated to on-chip memory for a selection of machine learning accelerators

DaDianNao [31]: 48%	Eyeriss [20]: 40–93%
EIE [28]: 93%	SCNN [27]: 57%
TPU [25] 35%	PuDianNao [29] 63%

As such a toolflow does not currently exist, in this text we concentrate on the manual construction of custom buffer hierarchies by the accelerator architect. This task has two major sub-components.

1. The selection of the appropriate buffering *idiom* (meaning caches, scratchpads, etc.) to accomplish the acceleration in the area/power/design-complexity budget.

2. The sizing of these buffers and their arrangement into a hierarchy that maximizes local accesses while minimizing stalls in the steady state.

To this end, we first describe a taxonomy of buffering approaches, describing both what has worked in the general-purpose computing community, and customized accelerator approaches. We use this taxonomy to discuss in depth the pros and cons of existing reusable buffer idioms such as *caches*, *scratchpads*, and *FIFOs*, with particular focus on their synchronization properties.

Furthermore, we describe *buffets* [32], an emerging storage idiom designed in the context of custom accelerators instead of general-purpose computers. Buffets are efficient and composable, and are not tied to any particular accelerator design or domain. We use the taxonomy classification to explore how buffets serve the needs of domain-specific accelerators in ways that caches, scratchpads, and FIFOs do not.

Finally, we discuss buffer composition into hierarchies, and considerations that go into sizing individual buffer levels. All in all we hope to convey to the reader that the buffering of a custom accelerator is a first-order architectural consideration—perhaps even more important than the datapaths themselves.

3.2   CLASSIFYING BUFFERING APPROACHES

The ideal accelerator buffer hierarchy would accomplish the following at minimal cost.

- Proactively transfer exactly the data that will be referenced in the future.

- Maximize the number of accesses to data in the smallest, fastest and most energy-efficient buffer.

- Stage data at the least-upper-bound buffer between sharers in the hierarchy.

- Overlap the fill of the next data tile with the consumption of the current data tile.

- Simultaneously broadcast (or multi-cast) the result of a buffer access to all consumers of the accessed data.

- Synchronize data availability precisely and cheaply without active polling.

- Remove data exactly when it is no longer needed.

In order to structure our thinking of existing reusable buffer idioms, Figure 3.1 presents a classification of deployment scenarios along two axes. At a high level, the implicit/explicit distinction refers to the level of workload knowledge that can be leveraged to control staging buffer decisions, while the coupled/decoupled axis refers to whether memory responses and requests are round-trip or feed-forward. We now present a detailed discussion of this taxonomy, and establish why these buffering schemes and discuss the pros and cons of each approach compared to the above ideals. Table 3.3 presents a comparative summary of the major points covered throughout the section.

3.2.1   IMPLICIT VS. EXPLICIT ORCHESTRATION

In the general-purpose computing community, caches (Table 3.3A) have served admirably as a reusable, modular buffer abstraction based on load/store operations. Algorithm 3.1 shows a typical interface for the cache buffer idiom. Although the engineering and area costs for a given cache hierarchy may be quite high, the effort is often amortized across several design points with re-parameterization. Caches have several desirable properties, such as composing invisibly into hierarchies. Memory-level parallelism—both multiple outstanding fills, as well as concurrency between fills and accesses to current contents—can be achieved using well-studied additional hardware (often called *lockup-free* cache structures).

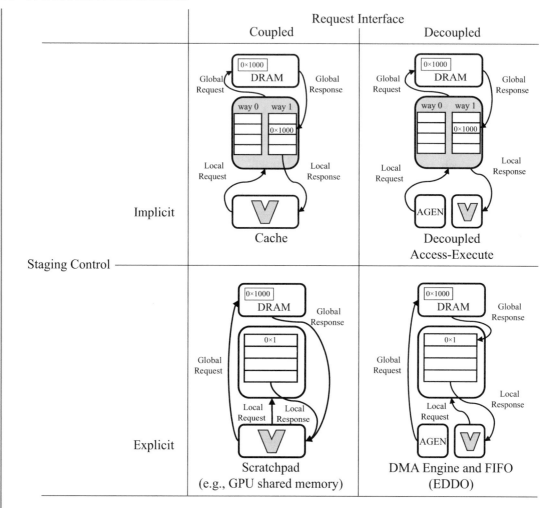

Figure 3.1: Taxonomy of data orchestration approaches, as used in typical deployment scenarios.

We say that caches perform *implicit* data orchestration as the load request initiator does not directly control the cache hierarchy's decisions about whether the response data is retained at any given level of the storage hierarchy, nor when it is removed. (In Figure 3.1 this is represented by the global request/response being shielded from the datapath.) Heuristic replacement policies are advantageous in general-purpose CPUs because they are workload agnostic.[1] On the other hand, for domain-specific accelerators, the area and energy overheads for features like

[1]As many programmers care more about optimization than portability, they often reverse engineer the details of the cache hierarchy and replacement policy to try to explicitly manipulate them. This is an indication that architects could provide more officially supported explicit data orchestration features in general-purpose processors.

Table 3.3: Summary of properties of traditional data orchestration approaches in typical deployment scenarios. Shaded cells indicate undesirable properties for domain-specific accelerators (although they may be acceptable for general-purpose architectures).

	(A) Datapath + Cache (*Implicit, Coupled*)	(B) Datapath + Scratchpad (*Explicit, Coupled*)	(C) D.A.E. Dpaths + Cache (*Implicit, Decoupled*)	(D) DMA + FIFO + Dpath (*Explicit, Decoupled*)	(E) DMA + Buffet + Dpath (*Explicit, Decoupled*)
Buffer non-RAM area	High	Low	High	Low	Low
Buffer access energy	High	Low	High	Low	Low
Placement policy	Workload-agnostic	Workload-controlled	Workload-agnostic	Workload-controlled	Workload-controlled
Achieving multiple fills in flight	Complex (lockup-free structures)	Complex (unrolling, multi-thread.)	Complex (lockup-free structures)	Straightforward (credit scheme)	Straightforward (credit scheme)
Achieving overlapped fill and access	Complex (static req. pipelining)	Complex (static req. pipelining)	Straightforward (dynamic rate matching)	Straightforward (dynamic rate matching)	Straightforward (dynamic rate matching)
Hierarchically composable	Yes	No	Yes	Yes	Yes
Landing zone hold time	Round-trip	Round-trip	Hop-to-hop	Hop-to-hop	Hop-to-hop
Access multicast	Dynamic coalescing	Dynamic coalescing	Workload-controlled	Workload-controlled	Workload-controlled
Data availability synchronization	Encapsulated (load-to-use)	Encapsulated (load-to-use)	Out-of-band (suplemental queue)	Encapsulated (peek stalling)	Encapsulated (read stalling)
Access order	Arbitrary	Arbitrary	Arbitrary	Strict FIFO	Arbitrary
In-place updates	Yes	Yes	Yes	No	Yes
Removal policy	Workload-agnostic	Workload-controlled	Workload-agnostic	Strict FIFO (or clear all)	Workload-controlled

tag matches and associative sets are currently considered unacceptably high. It is notable that no contemporary commercial machine-learning ASICs incorporate caches.

One alternative is to use *scratchpads* (Table 3.3B), which expose an address range of a particular staging buffer for loads/stores, thereby enabling *explicit* and precise control over the orchestration. (In Figure 3.1 this is represented by the datapath managing both local and global

---

Algorithm 3.1 Pseudo-code for a typical cache interface that supporting the return of hits while misses are outstanding (so-called lockup-free).

        ▷ Note: GLOBAL_ADDR datatype is independent of Cache size
        ▷ The size parameter is purposely not exposed via interface operations

module LockupFreeCache <UINT size, UINT width>:

    using DATA = BIT<width>

    interface Read:

        VOID ReadReq(GLOBAL_ADDR a)
        TUPLE<GLOBAL_ADDR, BOOL, DATA> > HitRsp()   ▷ True indicates Hit
        TUPLE<GLOBAL_ADDR, DATA> MissRsp()

                ▷ Read interface used for:
                ▷ Accessing staged data
                ▷ Staging new data (implicitly)
                ▷ Unstaging existing data (implicitly)

    interface Write:

        VOID WriteReq(GLOBAL_ADDR a, DATA d)
        VOID WriteAck()   ▷ Acknowledgements returned in order
                ▷ Write interface used for:
                ▷ Updating staged data
                ▷ Staging new data (implicitly)
                ▷ Un-Staging existing data (implicitly)

---

request/response.) Algorithm 3.2 shows the interface details, for contrasting with caches. A GPU's *shared memory* scratchpad [33] is the most widespread contemporary example of this idiom for explicit data orchestration. The size and address range of the scratchpad is exposed architecturally, and the transfer of data into and out of the scratchpad is managed via explicit instructions. While scratchpads avoid the hardware overheads of caches, extracting memory parallelism—both across fills and overlapping fills and accesses—is tedious and error-prone,[2] and as a result they are difficult to compose into hierarchies.

3.2.2   COUPLED VS. DECOUPLED ORCHESTRATION

Caches and scratchpads both use a load/store paradigm where the initiator of the request also receives the response. We call this a *coupled* staging of data, reflected in the left column of Figure 3.1. With this setup, synchronization between data demand and data availability is efficient and intuitive—the requester is notified when the corresponding response returns (load-to-use). The disadvantage to this approach is that it complicates overlapping the fill and access of data

---

[2]GPU shared memory is paired with high multi-threading and loop unrolling to offset these problems, but this complexity is considered unacceptable for fixed-function accelerators.

Algorithm 3.2 Pseudo-code for a typical scratchpad interface.

▷ Note: local address datatype is dependent on size parameter and therefore is exposed via interface types
▷ All out-of-range considerations omitted for brevity

module Scratchpad <UINT size, UINT width>:

    using LOCAL_ADDR = BIT<Log2<size> >
    using DATA = BIT<width>

    interface Read:

        VOID ReadReq(LOCAL_ADDR a)
        DATA ReadRsp()   ▷ Responses returned in order
                                  ▷ Read interface used for:
                                  ▷ Accessing staged data
        ▷ (Requestor manages whether any address contains live staged data)

    interface Write:

        VOID WriteReq(LOCAL_ADDR a, DATA d)
        VOID WriteAck()   ▷ Acknowledgements returned in order
                                  ▷ Write interface used for:
                                ▷ Updating staged data
                          ▷ Staging new data (explicitly)
           ▷ Un-Staging existing data (i.e., over-writing)

---

tiles (i.e., double-buffering) as the single requester/consumer must alternate between requesting and consuming responses. Additionally, a "landing zone" for the incoming data tile must be held reserved for the entire round-trip load latency, which increases pressure on RAM resources that could otherwise be used for larger tile sizes.

The alternative is to *decouple* the load request initiator from the response receiver. (In Figure 3.1 this is represented by the request/response arrows going to different modules.) In this setup, a separate hardware module (e.g., a DMA engine, or *address generator* (AGEN)) is responsible for *pushing* data into one or more functional units' staging buffers.[3] To tolerate latency, these are often double-buffered and hence sometimes referred to as ping-pong buffers [34, 35]. The main advantage to this approach is that the requester can run at its own rate, and can multicast data to multiple simultaneous consumers. Additionally, the feedforward nature of the pipeline means that the tile landing zone only needs to be reserved proportional to the latency between adjacent levels of the hierarchy, rather than the entire hierarchy traversal round-trip, allowing for increased utilization of equivalent RAM. Finally, this approach often can transmit large blocks of data, i.e., bulk transfers, which are more efficient than small requests, which must dynamically re-coalesce accesses to the same memory line.

---

[3]Cache pre-fetching can be considered an example of decoupling. Consideration of this large body of work is beyond the scope of this text.

Algorithm 3.3 Pseudo-code for a typical FIFO interface, including credit path to producer.

▷ Note: The size parameter is purposely not exposed in the Peek/Pop interface operations or types
▷ Ideally, consuming data from a FIFO is the same regardless of size

module FIFO <UINT size, UINT width>:

    using CREDIT = BIT<Log2<size> >
    using DATA = BIT<width>

    interface Push:

        CREDIT GetCredits()   ▷ 5 credits means bulk Push() (at least) 5 more times before FIFO is full
        VOID Push(DATA d)

                                                ▷ Push interface used for:
                                              ▷ Staging new data
                                      ▷ (Order is managed by the producer)

    interface Peek:

        DATA Peek() State

                                                ▷ Peek interface used for:
                                ▷ Accessing staged data (oldest element)

    interface Pop:

        VOID Pop()
        VOID Clear() State

                                                ▷ Pop interface used for:
                                          ▷ Un-Staging existing data

---

This separate producer/consumer approach is similar to Smith's [36] *decoupled access-execute* (DAE) style of general-purpose computing architecture (Table 3.3C). In a DAE organization two processors are connected by a hardware queue. The access processor is responsible for performing all address calculations and generating loads—analogous to the DMA engine. Load responses are passed to the execute processor—analogous to an accelerator's functional units and their local staging buffers. DAE improves parallelism and reduces the critical paths of instructions while allowing both processors to compute at their natural rate. However, classical DAE does not explicitly control data orchestration buffers—decisions about staging data are still managed by the cache hierarchy, thus Figure 3.1 categorizes DAE as implicit decoupled.

We advocate that the *explicit decoupled* data orchestration (EDDO) approach best matches the needs of domain-specific accelerators, as it allows for the best opportunities to leverage static workload knowledge combined with efficient, high-performance hardware. Hardware FIFOs [37, 38] (Table 3.3D) are one traditional reusable EDDO staging buffer organization. Note how the FIFO's interface (Algorithm 3.3) includes a path for synchronizing a remote producer that fills the buffer. The advantages are that FIFOs cleanly encapsulate synchronization via head and tail pointers, and are easily hierarchically composable. However, in practice, FIFOs are not flexible enough to meet the needs of modern accelerators, which often require random access within an active window or tile of data [20, 39–41]. Additionally, for data types such as the partial sums in a convolutional neural net, staged data must be modified several times in place before being drained [20, 42]. This is not possible in single write-port FIFOs without costly re-circulation. Thus, the goal of accelerator buffer architects is to combine the efficient hardware of FIFOs with the flexibility of scratchpads (Table 3.3E).

3.2.3   SYNCHRONIZATION CONCERNS

In a decoupled system, the timing of loading new tiles is a critical correctness component, as initiating a transfer too early could overwrite live data, and a transfer that is too late results in efficiency loss. Some accelerators use *systolic* approaches to bound tile processing time—essentially removing the need for synchronization hardware beyond simple counters. However, these time bounds often become overly conservative in realistic systems that have non-deterministic latencies involving off-chip accesses, arbitrated networks-on-chip, or functional units whose processing time involves conditional execution. Therefore, this text does not rely on them.

Instead, this chapter assumes that accelerators are built using a standard ready-valid (Rdy/Vld) micro-protocol for pipeline flow control as in SystemC compilation [43, 44]. For example, a FIFO could assert Pop.Rdy when it is non-empty, and dequeue the oldest element when Pop.Vld is asserted. Without loss of generality, techniques presented here can be applied using other micro-protocols such as Try/Ack. For convenience, we present an operation-centric interface [45] and omit the micro-protocol. For example, with FIFOs the following pseudo-code:

```
qC.Push(qA.Pop() + qB.Pop())
```

is shorthand for "when qC is asserting Push.Rdy (not full) and qA and qB are asserting Pop.Rdy (not empty), perform an addition and assert all Vlds." We refer to the micro-protocol details as appropriate for clarity in cases where this shorthand is not sufficient.

While circuit-level micro-protocols form the necessary groundwork for adjacent stages of a properly synchronized accelerator, they are not a complete solution for data orchestration, which is a higher-level concern spanning remote producers and consumers. In the next section, we present an EDDO storage idiom that encapsulates fine-grained synchronization within the staging buffer operations themselves. Encapsulation increases composability and re-usability, allowing accelerator architects to cheaply leverage the benefits of EDDO for multiple application domains.

3.3   THE BUFFET STORAGE IDIOM

Figure 3.2 depicts the data orchestration model of a *buffet*. In reference to the taxonomy found in Figure 3.1, buffets fall into the EDDO quadrant, as data movement is explicit and they use decoupled fill engines. The buffet approach expands the decoupling farther than traditional DMA

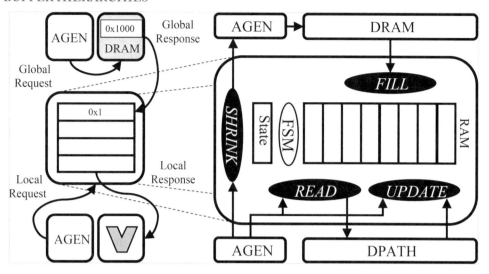

Figure 3.2: Buffet classification and operations.

setups by also using decoupled address generators to iterate over staged data. In Section 3.4 we discuss the composability benefits of this organization.

Algorithm 3.4 depicts a buffet's operational interface, which keeps the backward synchronization with a producer via credits, but combines it with the general ability of scratchpads to access and update all lines of the staged tile. Unlike scratchpads, these accesses are synchronized with the consumer, and stall appropriately when attempting to access regions of a tile that are not yet present.

Within the buffet, a finite-state machine controls the four fundamental storage operations: Fill(data), Read(index), Update(index, data), and Shrink(num). Upper-level modules in the hierarchy (e.g., DRAM) initiate Fill operations and move new data into the buffet, while lower-level modules (e.g., datapath) work with data in the buffet using Read and Update operations. Finally, Shrink operations remove data from the window.[4] Thus, the lifetime of a piece of data in the staging buffer can be described with the following regular expression:

$$Fill \rightarrow (Read \rightarrow Update?)^* \rightarrow Read \rightarrow Shrink. \tag{3.1}$$

Unlike scratchpad loads and stores, these operations encapsulate synchronization. We now present a detailed description of buffets' internal logic to arbitrate stalls.

---

[4]We choose the name buffet due to similarities with actual restaurants, where waiters bring out new dishes (fill) which are repeatedly iterated over by diners (read) until a course change (shrink). Of course, in restaurants diners are not allowed to return modified dishes to the buffet (update), due to food safety concerns!

Algorithm 3.4 Pseudo-code for a buffet interface, which can be seen as combining features of scratchpads and FIFOs.

      ▷ Note: The buffer size parameter is explicitly exposed to both the producer (like a FIFO) and consumer (like a scratchpad)

module Buffet <UINT size, UINT width>:

    using CREDIT = BIT<Log2<size> >
    using LOCAL_OFFSET = BIT<Log2<size> >
    using DATA = BIT<width>

    interface Push: ▷ Resembles a FIFO

        CREDIT GetCredits()   ▷ 5 credits means bulk Push() (at least) 5 more times before buffet is full
        VOID Push(DATA d)
                                        ▷ Push interface used for:
                                        ▷ Staging new data
                                        ▷ (Order is managed by the producer)

    interface Read: ▷ Resembles a scratchpad

                  ▷ In contrast to scratchpads the parameter is an offset relative to the oldest element
                                ▷ e.g., ReadReq(0, 0) == FIFO::Peek()
        VOID ReadReq(LOCAL_OFFSET o, BOOL will_update)
        DATA ReadRsp()
                                      ▷ Responses returned in order
                                    ▷ Read interface used for:
                                         ▷ Accessing staged data

    interface Write: ▷ Resembles a scratchpad

                        ▷ This interface only checked if a previous ReadReq() set will_update == 1
        VOID WriteReq(LOCAL_OFFSET o, DATA d)
        VOID WriteAck()
                                      ▷ Acknowledgments returned in order
                                    ▷ Write interface used for:
                                      ▷ Updating existing data

    interface Shrink: ▷ Resembles a FIFO

                      ▷ Note: Shrink(1) == FIFO::Pop() and Shrink(size) == FIFO::Clear()
        VOID Shrink(LOCAL_OFFSET count)
                                    ▷ Shrink interface used for:
                                    ▷ Un-staging existing data

3.3.1   BUFFET OPERATIONAL BEHAVIOR

Figure 3.3 presents a detailed architectural diagram of a buffet, and Algorithm 3.5 shows specific behavior. Newly transferred data is installed into the RAM via the Fill logic, labeled ❶. This path closely matches a conventional FIFO—no remote address accompanies the data, and placement is based on local address generation in the order it is received. (This is not a fundamental restriction, but the complexities of un-ordered Fill are beyond the scope of this chapter.) Un-

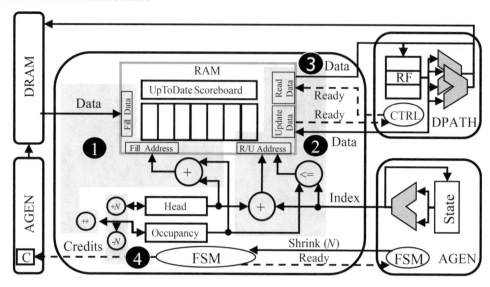

Figure 3.3: Buffet implementation details.

like a FIFO, the Read ❷ request includes an externally-provided index, allowing data to be read in a different order than it is received. The index is relative to the window, so 0 represents the oldest installed datum in the staging buffer. It is not legal for the index to exceed the physical size of the RAM.

Logic ensures that reading a position outside the active window will stall until the data arrives, similar reading an empty FIFO. However, the presence of this index means that buffet read stalling must be handled differently than traditional FIFOs, such as ones generated by commercial tools [37, 38]. In these circuits Pop.Vld is asserted whenever the FIFO is non-empty. In buffets, the presence of the requested data is a function of the index. Therefore, we use a separate *read response* path, represented by the "read_rsp_out.Send()" operation. ReadRsp.Vld is only asserted when the requested data has been filled, as in caches. For simplicity, we present read responses returning in the same order as requests—sufficient for the needs of most accelerators—but this is not fundamental (as in caches). In scenarios where blocking is harmful, a supplemental non-blocking Check(index) operation can be added to test if a certain index is in range.

Beyond indexed reads, a significant distinction from a FIFO is that data elements within the active window can be modified in-place, which we call the Update ❸ path. Internal logic stalls the modification of the RAM until both an index and a data element are asserting Vld. Thus, the index generation FSM and datapath can produce at different rates, and the system can tolerate dynamic timing variation. Semantically distinguishing RAM writes into Fill and Update operations raises the level of abstraction, allowing buffets to use customized synchronization logic for each case, as discussed below.

Algorithm 3.5 Buffet Operational Details

$\triangleright$ Initialize a buffet

function INIT(sz)
    head = 0;
    occupancy = 0;
    size = sz;
    credit_out.Send(size);

$\triangleright$ Emplace new data for staging.

function FILL(data)
    slot = (head + occupancy) % size;
    buffer[slot] = data;
    SetUpToDate(slot, true);
    occupancy++;

$\triangleright$ Iterate over staged data.

function READ(index, will_update)
    slot = head + index % size;
    wait_until(index < occupancy && IsUpToDate(slot));
    read_rsp_out.Send(buffer[slot]);
    SetUpToDate(slot, !will_update);

$\triangleright$ Update previously read locations.

function UPDATE(index, data)
    slot = (head + index) % size;
    buffer[slot] = data;
    SetUpToDate(slot, true);

$\triangleright$ Unstage data and free room for more Fills.

function SHRINK(num)
    wait_until(num < occupancy);
    head = (head + num) % size;
    occupancy = occupancy - num;
    credit_out.Send(num);

Finally, the `Shrink` path ❹ depicts the logic for removing staged data from the buffet. This operation takes a size parameter and removes that many elements from the active window. This operation simply updates internal scoreboarding—no data movement occurs. A credit is released to the `Fill` address generator indicating room for another bulk transfer. Accelerators that modify in-place data should drain it via the standard `Read` path before invoking `Shrink`. We advocate that the same index generation FSM that iterates staged data should generate calls to `Shrink`, removing the need for explicit stall logic between index-based read requests and re-basing of the oldest element in the active window. This arrangement is not fundamental, but for simplicity of synchronization we use this presentation for the remainder of this chapter.

3.3.2   BUFFET SYNCHRONIZATION DETAILS

We leverage the semantics of buffet operations to provide fine-grained synchronization with minimal logic and stalls. The cases where explicit hardware synchronization is required are represented by the "wait_until" calls in Algorithm 3.5. Many other cases of synchronization can be handled without the need for explicit hardware support because of the order of operations

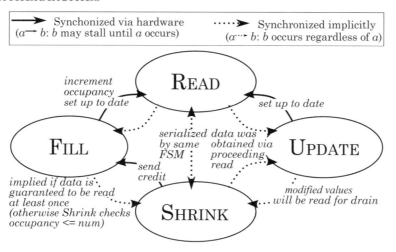

Figure 3.4: Operation synchronization relationships.

shown previously in Equation (3.1), as shown in Figure 3.4. For example, because modified values are always Read before Shrink, we do not need synchronization logic between Update and Shrink—they are transitively synchronized through Read.

Because modification of data by the datapath can take notable latency, buffets add RAW-hazard checks, which we present abstractly as an "UpToDate" scoreboard with perfect knowledge. In practice, any physically efficient tracker can be used, including imprecise hashing schemes. For simplicity, we present will_update as a parameter to Read that indicates that the datapath will modify the currently staged value—many alternative interfacing paradigms can be conceived of, including Lock() methods. If a subsequent Read requests an index that is undergoing modification, the response is stalled from returning—indistinguishable from reading an index that has not yet been filled. Fill writes do not need to perform this check. Thus, we can improve energy efficiency and performance using higher-level knowledge.

Encapsulating hazard detection inside the buffering interface removes the need for engineering custom stall logic on a per-deployment basis. To meet the efficiency demands of domain-specific accelerators, buffets provide options for design-time customization. If the architects can prove that no RAW hazard is possible, then the RAW hazard detection logic can be statically removed via a parameter. Similarly, if Fills are proven to be mutually exclusive with Updates then they can share a write port rather than using a more expensive RAM with multiple write ports. Finally, for buffets that hold read-only data the entire Update path can be removed. The goal is to allow designers to quickly construct a functionally correct orchestration hierarchy that can serve as a starting point for optimization refinements.

Another case for design-time optimization is the synchronization between Shrink and Fill. Algorithm 3.5 conservatively uses explicit synchronization for Shrink, e.g.,

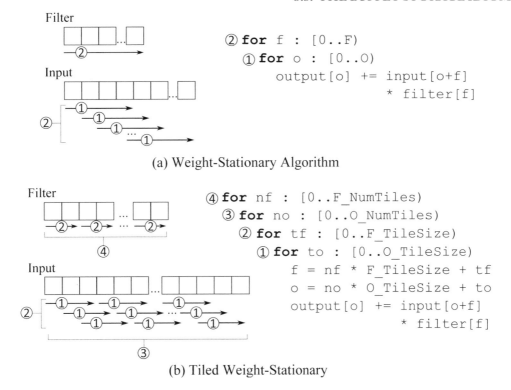

(a) Weight-Stationary Algorithm

(b) Tiled Weight-Stationary

Figure 3.5: 1D Convolution example orchestration.

"wait_until(num < occupancy)." This logic can be removed at design time if the architects can prove each staged data element will be read at least once. Figure 3.4 depicts this as implicitly synchronized, as we expect this to be the more common case. Synchronization with the external fill request generator is handled via backward credit flow—similar to a NoC protocol, discussed in detail in Section 3.4.

3.3.3   EXAMPLE ORCHESTRATION WITH BUFFETS

Figure 3.5A shows an example 1D convolution that demonstrates the use of buffets for EDDO. We purposely choose a weight-stationary dataflow [20] which involves re-loading partial sums several times until the final sum is produced to demonstrate all buffet features. These same principles generalize to other dataflows of full CNNs and other kernels.

The baseline formulation maximizes reuse of filter weights with only a single on-chip register, but it must resort to expensive off-chip accesses for inputs and outputs (if 2× the size of O is larger than on-chip buffering). Figure 3.5B shows a more realistic tiled formulation. This

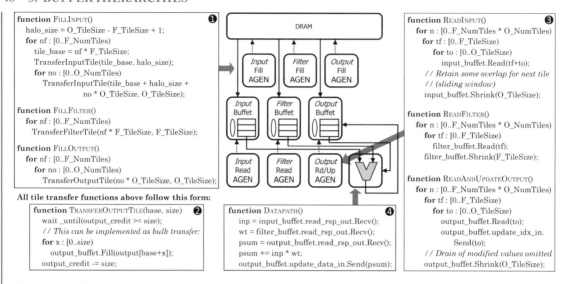

Figure 3.6: Basic accelerator of Figure 3.5 using buffets, with pseudo-code describing the EDDO data transfers.

introduces some re-reads of weights, but significantly reduces reloads of inputs and outputs as the tiles can be held resident on-chip.

Figure 3.6 shows a straightforward accelerator implementation of this algorithm using buffets. This accelerator uses separate buffers per datatype—similar to an instance of Eyeriss [20] with a single processing element and a different dataflow. Following the EDDO principles, separate address generation FSMs (labeled ❶) generate requests to the DRAM that install data into the staging buffers. The weight-stationary dataflow means that the weights are staged once while input and output tiles are re-staged per weight tile. The "wait_until" in the transfer FSMs ❷ represents blocking until sufficient credit is available. The backward path that increments the credit count is not shown.

Looking in-depth at the Read FSMs ❸, we highlight several points that distinguish buffets from FIFOs. First, both the input and output buffet are performing window-based access relative to the oldest element. Second, the size-based shrink gives more control over data liveness—the input is reading a tile volume larger than its shrink, and so represents a sliding window. Finally, the Output buffet is being used to perform in-place updates of staged sums. Each sum is modified F_TileSize times per tile. The buffet's internal scoreboarding ensures that subsequent Reads will block on previous Updates—a scenario which could occur if the multiply-accumulate datapath was implemented as pipeline with internal latency. As stated above, unnecessary scoreboard logic can be removed using design-time parameterization. In this algorithm, every partial-sum output that is read is also modified, so the accelerator can use the `ReadAndUpdateOutput` FSM to generate `Read` and `Update` indices. Other scenarios can require separate index generators for these two classes.

Furthermore, the `Read` FSMs reveal some key ways that buffets differ from traditional scratchpads. In a scratchpad, read requests return the current RAM value, so it is the responsibility of the iterator FSM to not issue a read request until it knows the desired data has been staged. Buffets' encapsulation of fine-grained synchronization means that no explicit checks are present in the `Read` FSM. The index generator issues `Read` operations at its natural rate—if the requested data is not available, then the `read_rsp.Vld` signals will stall the `Datapath` FSM ❹, as described in subsection 3.3.1. Furthermore, with scratchpads all accesses are done via absolute addresses into the underlying RAM, which places the burden of dividing the RAM into active and inactive regions onto the index generator. With buffets, the FSMs contain no explicit base address manipulation, nor wrapping-around of addresses relative to the size of the RAM. This hardware has not disappeared, but has been encapsulated inside the buffet, simplifying the creation of the index generation FSM to only repetition counts, bounds, and offsets.

This encapsulation of the modular arithmetic is important for another reason: offset-based indexing separates the size of the active tile from the size of the underlying storage which is pre-buffering future tiles. One way this manifests is that Figure 3.6 does not specify concrete sizes for the underlying RAM. If it is equal to the tile size, no pre-buffering will occur, as each buffet can only hold the active window. If it is twice the tile size, then the arrangement is equivalent to traditional double-buffering. It can also be set to any arbitrary constant. This flexibility has an important implication: the functionality of the FSMs in Figure 3.6 is unchanged across all these options, and do not need any alteration or re-verification if the underlying RAM size of the buffets is increased as part of design-space exploration. This will not affect correctness, only performance as more room is available to pre-fill tiles.

Designers often talk about double-, triple-, or even quad-buffering, but extra buffering should not be limited to multiples of tile size. By changing the Fill FSMs to transfer after receiving smaller credit totals such as 1/4, or 1/8 tile (or even arbitrary absolute values unrelated to tile size) architects can determine the optimal buffering needed to tolerate the forward latency through the memory system, which is unlikely to be an exact multiple of tile size. Additionally, if some datapaths are farther from the memory buffets' encapsulation makes it straightforward to give them more buffering for extra latency tolerance without altering their logic. In effect, this approach makes the staging buffer RAM sizes a low-level micro-architectural feature that can be determined late in the design process based on underlying physical properties, rather than a first-order design consideration.

3.3.4   AUTOMATICALLY DERIVING CONFIGURATION

The EDDO approach requires the implementation of separate index generation engines per buffet. These iteration loops all relate to the original tiling and inter-tile schedule chosen by the architects using workload knowledge. As a specific example, the loop for `ReadFilter` in

Figure 3.6 removes the `O_TileSize` loop level because in the original algorithm (Figure 3.5) the variable o is not used to index the `filter` array. This implies that the same filter weight value is being held local to the datapath while the input and output changes with o—hence the name "weight stationary." Similarly, the sliding window on the input buffet is created by addition of the f offset into the `input` vector.

We believe that automatic derivation of control FSMs from a single tiling specification is feasible future work. This has been accomplished before for specific high-value domains [46, 47] but not in a generalized tool. The domain-specific nature of accelerators limits the data orchestration patterns that need to be supported by automatic generation solutions. Furthermore, architects of accelerators can always fall back on handcrafting the data flows if the tools fail. The design of such tools is beyond the scope of this chapter.

In summary, buffets' encapsulation of synchronization into the operational interface allows design to more productively occur at a higher level of abstraction. We view buffets as combining the best properties of FIFOs, N-buffered scratchpads (e.g., double- or triple-buffered, etc.), and custom sliding window buffers. In the next section we will demonstrate that these same features also enable composition of buffet hierarchies.

3.4   COMPOSITION OF BUFFER IDIOMS

No staging buffer can be considered entirely in isolation. Domain-specific accelerators require the ability to hierarchically stage and distribute data, which facilitates maximizing data reuse from small buffers physically co-located with functional units. This section describes the interaction, modularity, and composability of various accelerator buffer idioms.

3.4.1   BUFFER HIERARCHIES

Ideally, all buffer idioms would be able to be seamlessly arranged into hierarchies without the exact topology being exposed to the accelerator datapaths. This capability then allows the buffers to be designed separately and modularly from the datapaths themselves. As discussed previously in Section 3.2, and reflected in Table 3.3, caches, FIFOs, and buffets all have this property, while traditional scratchpads do not. This is due to the lack of synchronization encapsulated in the scratchpad. In a scratchpad hierarchy, each L2 or L3 scratchpad must have an owner—meaning hardware that stages and reads data locally. One potential solution is to reserve memory regions of the scratchpads themselves as semaphores—equivalent to parallel programming paradigms. But this results in expensive polling or memory monitors, defeating the simplicity of the scratchpad idiom (and requires reasoning about load/store ordering in a distributed memory model). The other alternative is to design a side-band synchronization between the owner FSMs, but this requires designing protocols that are on the same order of complexity as FIFO credit protocols or cache-coherence policies. After such a scheme is in place it is arguable that the result no longer should be classified as a scratchpad, but an entire new column in Table 3.3.

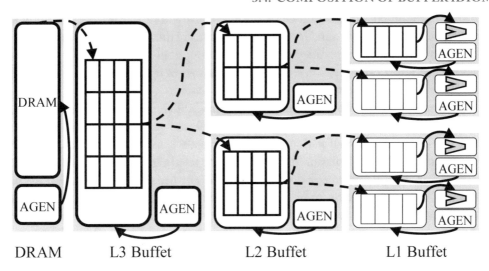

DRAM   L3 Buffet   L2 Buffet   L1 Buffet

Figure 3.7: An example buffet hierarchy.

Figure 3.7 shows how the credit protocol used by FIFOs and buffets allows those idioms to be seamlessly arranged into a hierarchy. We now walk through this example using buffets, as FIFOs can be thought of a subclass with limited address generation orderings. The upstream buffet's address generation FSM's Read operations take on a role similar to a traditional DMA engine, driving data to Fill the next buffet using the crediting mechanism presented previously. The downstream buffet's Read FSM blocks locally, without any external polling. Thus, serially composing N buffets requires $N + 1$ address generators, not $2N$. Larger tiles are staged, then broken down and distributed to lower levels for further staging and processing. We show data as flowing through all levels, but this is not fundamental—fills can bypass intermediate levels as appropriate. The data also need not be inclusive as upstream buffets can shrink away data before it is fully consumed downstream.

Although FIFOs and buffets use the same Fill crediting, an important distinction is that a hierarchy of buffets allows applications to fully exploit data reuse at each level in the hierarchy. Specifically, buffets' arbitrary iteration ordering means that data iteration order can be changed at each address generation level. This both allows interleaving of tile delivery across lower levels and altering the data layout in the memory dynamically so that each level may use an individually customized layout.

In any buffer hierarchy, backward paths are necessary to return modified data to larger, upper-level buffers for off-chip I/O. Traditional FIFOs have no support for in-place updates, so often neural network accelerators will use FIFOs for read-only datatypes such as inputs and weights, and scratchpads for partial sums. The buffet idiom encapsulates this using the Update functionality. Thus, the slot for the value is held in a non-up-to-date state by the upper-level buffet until the modified value is produced. As a result, `Update` traffic does need any kind of credit-check mechanism. Additionally, a major distinction from general-purpose caches is that there is no requirement that modified values pass through all levels of staging buffer during writeback. Accelerators can provision connectivity to transmit results from the datapath directly to higher levels, or to off-chip memory.

There is another significant distinction between Explicit and Implicit buffer hierarchies. In any Explicit hierarchy, both `Read` and `Update` requests are generated locally, any given index FSM only needs to concern itself with generating addresses sized to its local RAM. This means that low-bitwidth address calculation datapaths can be used rather than 32- or 64-bit address arithmetic. (Domain-specific accelerators generally do not use address translation for on-chip accesses, but it is not uncommon to use large-bitwidth addressing to ease system integration.) On the Coupled/Decoupled axis, buffer idioms that are decoupled never transmit addresses remotely, as address generation FSMs are co-located with the memory RAMs, which can reduce transfer bandwidth and energy.

No buffer hierarchy should ever make an assumption about being deployed with direct connections between buffer levels vs. being connected via a network-on-chip. In the latter case, there is extra synergy for Decoupled idioms, as the credit flow for forward `Fills` and use of `Update` for backward writeback of modified values act as a guarantee of deadlock freedom, as any injected data will be unconditionally drained—a so-called "consumption assumption." This guarantee can simplify network flow-control design and verification, and result in simpler hardware. For example, buffet-based accelerators may require fewer or no virtual channels for correctness—although they still may still be employed for routing or quality-of-service reasons. Furthermore, the FIFO's and/or buffet's `Fill` path may be directly attached to the output of a router port, essentially supplanting the need for separate egress buffer hardware.

3.4.2 SHARING FILLS VIA MULTICAST

If the transport substrate supports broadcast or multicast, either via dedicated wires or a network, then Decoupled idioms can leverage it with significantly less complexity. Coupled idioms such as caches and scratchpads generally do not have the capability to accept "unexpected" data fill responses—i.e., arrival of data that was not previously asked for by the buffer. Again, although adding this capability is possible, it results in a significant increase in complexity.

Multicast is a significant source of energy efficiency, as it means the output of one access to a large physical RAM can be delivered efficiently to all consumers. We advocate that accelerators leveraging static workload knowledge is more area- and energy-efficient than dynamically detecting and coalescing multiple requests to the same address in a cache or scratchpad hierarchy, and does not rely on multiple requests arriving within a limited time window.

To achieve multicast we use a straightforward extension of the credit scheme presented in Section 3.3. The `Fill` FSM's credit register is extended to a vector tracking credits of multiple target FIFOs/buffets. All downstream targets must have sufficient room before a transfer can be initiated, and credits are decremented from all targeted counters upon transmission. Finally, the backward credit path is supplemented with an ID field indicating from which FIFO/buffet the credit originated, which is used to increment the appropriate counter in the credit vector. An extension of this scheme uses run-time configurable routes and IDs allows the design of accelerators with dynamically reconfigurable buffet hierarchies.

3.4.3   SHARING PHYSICAL RAMS EFFICIENTLY

Within each level of the buffer hierarchy, an important decision that accelerator buffer architects must make is whether to use *partitioned* or *shared* memory structures between datatypes. For example, the Eyeriss neural network accelerator [20] uses distinct L1 RAMs for inputs, outputs, and weights. This results in the simplest possible hardware, but can result in under-utilization if not enough data of any particular tile is available. This fragmentation is doubly bad because it results in a missed opportunity cost—the ability to expand the tile size of one of the other datatypes into the empty entries. If this is possible, it both can increase reuse, and potentially reduce the total number of tiles.

On the other hand, sharing the same RAM pool between multiple datatypes can both avoid internal space fragmentation and decrease per-RAM overheads. To implement this feature, we supplement each logical buffer's bookkeeping with base and bound registers. In FIFOs and buffets, all increments to the head pointer are carried out modulo base and bound. Various multi-port and/or banking schemes can be used to maximize RAM efficiency by matching the required bandwidth across data types. For example, the accelerator previously presented in Figure 3.6 uses a weight stationary dataflow which means that new filter weights are only required every O_TileSize cycles, whereas new inputs and outputs are required every cycle. Therefore, it may be profitable for the filters to share the same RAM as either the inputs or outputs, but multiplexing the inputs and outputs themselves would reduce throughput more significantly.

Overall, RAM-sharing buffets represent a combination of design-time customization and runtime flexibility. However, the reconfiguration of buffer sizes requires both hardware area and design complexity, and may increase the energy cost per access. Whether buffers are shared or partitioned is a critical tradeoff for accelerator architects to focus on, and may result in heterogeneous topologies. For example, Eyeriss supplements the partitioned buffers at the L1 with a re-configurable shared buffer at the L2 [20].

3.4.4   EXAMPLE OF HIERARCHICAL ORCHESTRATION

We now revisit the 1D convolution example presented in subsection 3.3.3 in the context of a more realistic accelerator with parallel functional units, a multi-level buffet hierarchy, and physically shared RAM. Figure 3.8 shows an extension of the weight-stationary dataflow by spatially partitioning individual tiles across the separate functional units—represented by parallel-for loops. We use the term "partition" to refer to tiles that are executed by parallel hardware instances. As the number of partitions is a design-time parameter, any remaining mismatch

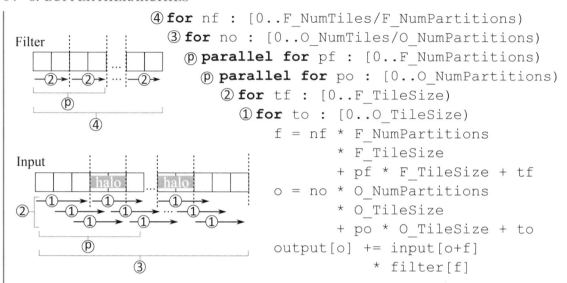

```
                    ④ for nf : [0..F_NumTiles/F_NumPartitions)
                  ③ for no : [0..O_NumTiles/O_NumPartitions)
                ⓟ parallel for pf : [0..F_NumPartitions)
                 ⓟ parallel for po : [0..O_NumPartitions)
               ② for tf : [0..F_TileSize)
              ① for to : [0..O_TileSize)
                  f = nf * F_NumPartitions
                          * F_TileSize
                          + pf * F_TileSize + tf
              o = no * O_NumPartitions
                          * O_TileSize
                          + po * O_TileSize + to
             output[o] += input[o+f]
                          * filter[f]
```

Figure 3.8: **Parallel partitioned version of** Figure 3.5.

between data size must be handled as passes in the outer loop. We omit edge cases from the algorithm.

Figure 3.9 shows an accelerator with F_NumPartitions =2 and O_NumPartitions=2. Buffets within a level share the same multi-banked RAM, which could potentially allow F_TileSize and O_TileSize to be configured at runtime, with the constraint that the total tile size across all data types size fits in the underlying RAM. This dataflow employs separate partitions updating the same partial sum, and so requires an additional datapath for reduction before updating DRAM. This path is not shown in the figure as it has no effect on the underlying orchestration patterns.

Despite the pedagogical nature of this accelerator, it has several interesting characteristics. First, each L2 buffet holds a disjoint filter tile, so the L2 filter buffets are filled via unicast from DRAM. The delivery of tiles can either be interleaved or serial, as determined by the iteration order of the Read index generation FSM. Partitioning filters means that inputs are broadcast from DRAM to L2, as the same input must be convolved against all weights (excepting the edges).

The L2 buffets' Read FSMs distribute the second level of spatial partitioning. Each L1 processes a different output tile against the same weights, therefore output tiles are now unicast. One the other hand, the filter tiles are now broadcast from the L2 to all connected L1s. For inputs, the tile is unicast, but the halo region depicted in Figure 3.8 can be multicast simultaneously to two L1 buffets. In the steady state, this savings is doubled because of overlap with both previous and next partitions.

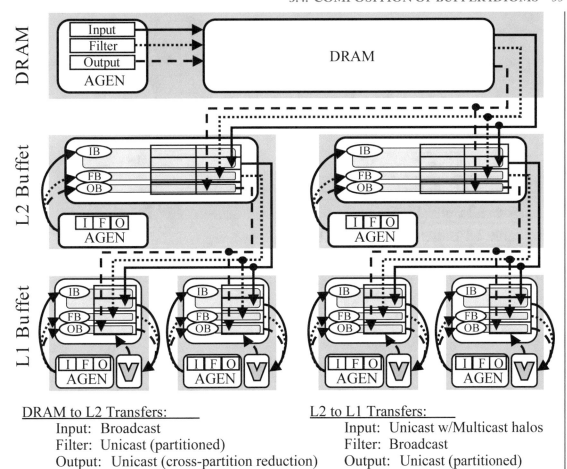

DRAM to L2 Transfers:
    Input: Broadcast
    Filter: Unicast (partitioned)
    Output: Unicast (cross-partition reduction)

L2 to L1 Transfers:
    Input: Unicast w/Multicast halos
    Filter: Broadcast
    Output: Unicast (partitioned)

Figure 3.9: Example accelerator derived from the EDDO in Figure 3.8 with multiple levels of buffets sharing RAMs. For simplicity, paths for cross-partition reduction of partial sums are not shown.

Multicast can represent a significant savings. Figure 3.10 shows an analysis of two Figure 3.9-style accelerators that differ only in support of multicast. No standard 1D convolutional benchmark exists, so we approximated data set sizes by examining the work to produce one output channel of VGG-Net16 [3], convolutional layer 13, and projecting the data ratios down into one dimension, giving an input size 21.78× larger than the filter. The multicast version reduces traffic between staging buffers to 58.9% of the unicast-only arrangement, and therefore performs only 58.9% of RAM accesses as well.

`F_NumPartitions`	10
`O_NumParitions`	10
Total Functional Units	100
`F_TileSize`	256
`O_TileSize`	1024
Indiv. L2 Buffet Size	45.4 KB
Total L2 Buffet Size	455.0 KB
Indiv. L1 Buffet Size	5.0 KB
Total L1 Buffet Size	500.0 KB
Input size	1.64 MB
Filter size	74 KB
Input: Filter ratio	21.78

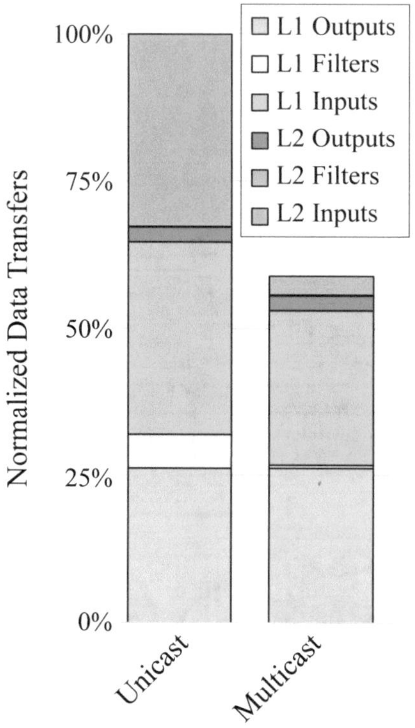

Figure 3.10: Analysis of the impact of enabling multicast on a Figure 3.9-style accelerator.

3.5 OTHER RELEVANT BUFFERING IDIOMS FOR ACCELERATORS

Many accelerators use a customized buffering solution based on their application [20, 27, 31, 48, 49]. While these proprietary buffering designs are efficient, it is difficult to generalize their approaches to future accelerators. In this chapter we have focused on reusable, accelerator-independent buffer idioms. In the long term, the accelerator community needs a general reusable accelerator storage design to provide high performance with minimal effort. The goal of buffets is provide this in the EDDO context. Table 3.4 summarizes other notable proposals for general storage solutions for accelerators.

Jenga [50] treats the cache SRAM banks as a general pool that can be used to create distributed virtual cache hierarchies that are targeted to given application. This solution uses the implicit coupled idioms to provide a cache based solution. This approach provides a more efficient solution than a traditional cache but still has more overhead than an explicitly managed memory architecture.

Table 3.4: Categorizing storage idioms for accelerators

Jenga [50]	Implicit	Coupled
DeSC [51]	Explicit	Decoupled
PDAE Cache Prefetch [39]	Implicit	Coupled
PDAE DAE [39]	Implicit	Decoupled
Stash [52]	Implicit	Coupled
Accelerator Store [53]	Explicit	Coupled
Patch Memory [54]	Implicit	Coupled
LEAP Scratchpads [55]	Implicit	Coupled
CoRAM [56]	Explicit	Coupled
Stream dataflow [57]	Explicit	Decoupled

Decoupled Supply-Compute (DeSC) [51] uses a traditional DAE approach to divide memory accesses instructions from compute instructions and place them on separate datapaths, connected by a queue. However, the queue size in DESC is not architecturally visible, and therefore cannot be used as an explicit target for determining data tiling size. Additionally, DeSC uses a traditional DAE FIFO queue which prevents random iteration without transferring the data to the L0 registers thus limiting access order and preventing in-place updates. It also has a high non-RAM overhead associated with having a full OoO processor as the access engine, although this is not fundamental as accelerator architects could could replace this with a specialized FSM.

Prefetching with decouple access-execute (PDAE) [39] has two hierarchies for data orchestration. PDAE uses a prefetching cache for accelerators with regular access patterns and a traditional decoupled-access-execute approach for systems with irregular access patterns. The prefetching cache has traditional cache overheads while the DAE approach has similar iteration restrictions as DeSC.

Stash [52] is a data staging scheme for accelerators that makes scratchpads more like caches. It unifies scratchpads into the global address space like a cache, but uses a specific user-provided translation function to fill the scratchpad on a miss. Translation reduces the number of explicit operations required to fill the scratchpad through the creation of a load-and-store-scratchpad. Stash does not use a decoupled fill approach, and explicit double-buffering is still required for concurrency.

The Accelerator Store [53] is a shared memory scratchpad that allows multiple logical buffers to be mapped onto a single scratchpad, similar to buffets. The logical buffers can be configured in FIFO, random-access, or a hybrid mode. However, RAMs are accessed by coupled datapath load/store (or get/put for FIFO mode) operations, without any dedicated index generators and synchronized using bulk interrupts.

Patch Memory [54] is a domain-specific scratchpad specifically designed for data tiling in image processing accelerators. Users define patch parameters and dataflow order between patches, and data is filled in a DAE style. Iterations into the patch use coupled loads and stores from the datapath without DAE. Patch memory is not hierarchically composable.

LEAP scratchpads [55] are a memory architecture for FPGAs that allow users to define logically separate scratchpads that are mapped onto a unified cache structure with tags and traditional fill-on-load behavior. Concurrent accesses to other logical scratchpads are possible while one scratchpad is filling. Based on the caches, they are hierarchically composable.

CoRAM [56] is a memory architecture for FPGAs used as accelerators. CoRAM defines a set of operations that is used to construct custom fill engines and scratchpads that are programmed into the FPGA's logic and block RAM. Fine-grained synchronization is possible directly through FPGA signals. However, CoRAM does not use DAE for iteration accesses, but rather a traditional load/store interface. Additionally, no indication is given of an ability to compose CoRAM hierarchically, although this may be possible given the generality of the CoRAM operations.

Stream dataflow [57] is an architecture and programming model based on streams and CGRAs. The data is stored locally in a scratchpad and streamed into a dataflow CGRA for efficient acceleration. The architecture uses an address generators for accessing the memory system and scratchpad. However, the synchronization of the accesses requires the use of explicit barriers. This need of barrier prevents overlapping fill and access. The stream-dataflow architecture does not support in-place updates due to the flow of the data through the system. Furthermore, their memory system was not designed to be composable, although this could be accomplished as future work.

3.6   RESEARCH NEEDS FOR ACCELERATOR BUFFER HIERARCHIES

The domain-specific accelerator era of heterogeneous computing is poised to tackle key problems in compute through the integration of application specific engines. However, achieving this efficiency requires tremendous effort to design and verify a wide variety of accelerators. In this chapter we identified commonalities in the efficient transfer and staging of data that can be leveraged to generate a design-effort and area-efficient accelerator memory system. But no such automated toolflow to leverage these properties exists today.

The Implict/Explicit and Coupled/Decoupled taxonomy reveals why reusable idioms that have worked well in the general-purpose computing space fall short for accelerators. EDDO is emerging as the tactic that best allows accelerator architects to leverage their static workload knowledge for efficiency. However, it is crucial to realize that FIFOs—the traditional reusable buffering idiom in this space—fail to serve the needs of accelerators. The buffet idiom is developed specifically to address this, but currently all deployment of buffets is manual, including hierarchical composition, multicast, and RAM sharing.

A key component that requires focus from the community would be the development of a domain-agnostic accelerator toolflow that can generate orchestration FSMs and hardware configurations. We believe that run-time reconfiguration of orchestration patterns can be added into this toolflow as well. This feature allows data orchestration to be tuned to the particular sizes and ratios of the current data-set, rather than settling for an average-case approach, which has shown to be advantageous for individual accelerators [42, 58–60]. We hope that accelerator-independent EDDO abstractions such as buffets help bring together the necessary engineering effort to create such a toolflow, rather than each accelerator creating their own tools.

3.7 SUMMARY

Accelerator buffer hierarchies are a rich sub-field worthy of large amounts of dedicated engineering and research effort. In this chapter we enumerated the constraints and goals dedicated domain-specific accelerators place on buffer hierarchies, and described efforts such as buffets that have arisen from these constraints. Overall, there is a large amount of future effort that needs to be put into dedicated buffering.

Note that increasing the capabilities and features in the buffer hierarchy is useless if the system around it is not flexible enough to adapt to these changes. Before we are able to design a complete accelerator we must discuss in-depth the data delivery networks that fill buffers and feed datapaths.

CHAPTER 4

Networks-on-Chip

As discussed so far, the massive parallelism opportunity of DNN algorithms has given rise to a suite of spatial architecture-based accelerators, which contain an array of hundreds of PEs. These accelerators aim to achieve high throughput by exploiting massive parallel computations over the PEs while keeping the cost-of-operation much lower than off-the-shelf components with the same compute budget. However, adding more compute elements in an accelerator is not sufficient to achieve high throughput because more compute elements require more operands read from the on-chip memory hierarchy, and delivered to each compute element. Similarly, the output activations generated by the PEs also need to be written back to the on-chip memory hierarchy for the next layer.

In this chapter we focus on on-chip data movement characteristics within DNN accelerators and explore network-on-chip (NoC) architectures to address them. We first discuss the various traffic movement patterns in typical DNN accelerators. Next, we present background in traditional NoC design and understand their limitations. Finally, we go over the design of NoCs specially crafted for the traffic movement patterns within DNN accelerators. We want to emphasize that the focus of this chapter is on the scale-up interconnect fabric *within* one accelerator, not a scale-out off-chip network connecting multiple accelerators [61].

4.1   COMMUNICATION PHASES

Communication within DNN accelerators involves the following two phases.

- **Distribution.** Distribution is communication from the global buffer to PEs, which delivers weights and input activation values to be used in each PE. Distribution may involve multiple unicasts, or a mix of unicasts and **multicasts**, depending on the dataflow approach and exposed spatial reuse opportunities. We define multicast factor as the number of delivered data (including duplicates) to each PE divided by the number of global buffer accesses. A multicast factor greater than one implies effective read bandwidth amplification from the global buffer. Multicasts may be implemented via an explicit fanout network (e.g., bus or tree) or via a store-and-forward mechanism.

- **Collection.** Collection is communication from PEs to the global buffer to deliver final output activations produced within the PE array. Collection requires multiple unicasts from one or more PEs (depending on how many unique outputs the array generates based on the mapping strategy) to the global buffer. Depending on the accelerator's dataflow, collection may include a ***reduction*** step for accumulating partial sums. In some dataflow/mappings, reduction is purely temporal within a PE. In others, it is done spatially through a linear reduce-and-forward mechanism or via reduction trees, or a hybrid of the two. If the mapping requires spatial reduction, but there is no spatial reduction support within the communication network, then partial sums from different PEs would need to be sent to the global buffer and then re-distributed to a PE for accumulation. Reduction of partial sums within the array (temporal or spatial or hybrid) leads to an effective decrease in write-bandwidth at the global buffer. From this point of view, spatial reduction in collection can be seen as directly analogous to multicast in distribution, with the major difference that the many-to-one communication tree also transforms (i.e., reduces) the data passing through.

4.2   TRADITIONAL NETWORKS-ON-CHIP

NoCs is the term used to describe the interconnection topology, routing, flow-control, and microarchitecture employed when connecting multiple on-chip PEs and memory banks together. This section provides a brief primer on traditional NoC architectures. We refer the readers to the Synthesis Lectures on *On-Chip Networks* for more details [62]. In the DNN accelerator context, we use the term "node" to refer to either a PE or a SRAM bank.

4.2.1   TOPOLOGY

There are two classes of NoCs: *application-specific* and *general-purpose*. Application-specific NoC topologies are designed in accordance with apriori knowledge of the application's communication graph [63] and are common in MPSoCs in the embedded domain. This is shown in Figure 4.1a. The bandwidth on different links can also be tuned independently to meet the expected traffic flow from each node. It is certainly possible to design a NoC topology that mirrors the connectivity of a specific DNN by mapping each neuron on a PE; however, this is infeasible for the following reasons. First, the number of neurons in modern DNNs are much larger than the number of PEs that can fit on-chip, thus requiring multiplexing of multiple layers with different connectivities over the same physical PEs and wires. Second, most DNN accelerators are designed to run more than one DNN, which again necessitates multiple possible DNN connectivities mapped over the physical wires. NoC topologies within DNNs thus need to be more general.

General-purpose NoCs are used extensively in many-core processors today. They employ regular topologies, with fixed link-widths. We describe the most common topologies next. They are illustrated in Figure 4.1b.

- **Bus:** A bus is the simplest NoC implementation where all nodes tap onto a shared wire. Only one node is allowed to send at a time, and all other nodes snoop. Thus, buses are excellent for implementing broadcasts and multicasts (i.e., spatial reuse during distribution). The key

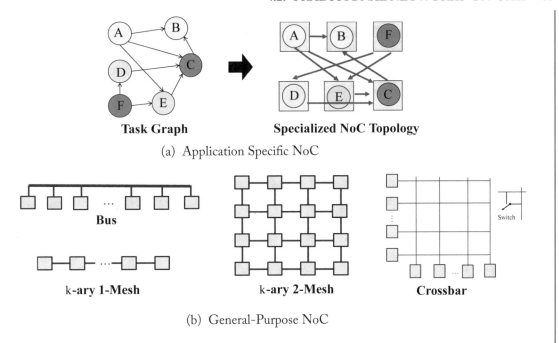

(a) Application Specific NoC

(b) General-Purpose NoC

Figure 4.1: Traditional NoCs used in ASICs and many-core chips today.

limitation of buses, however, is bandwidth, since there can only be one transmission at a time.

- **Crossbar:** A crossbar is a collection of tiny switches connected like a matrix. The rows are connected to the source nodes, and the columns are connected to the destinations. Crossbars allow non-blocking communication between any source-destination pair each cycle, as long as there is only one request for each destination. Crossbars are extremely valuable for throughput, but are not scalable due to area and power concerns.

- *k*-**ary 1-Mesh:** A *k*-ary 1-Mesh is a linear arrangement of *k* nodes with point-to-point wires (uni-directional or bi-directional) between neighbors. The effective bandwidth of such a topology is higher since every node can signal its neighbor every cycle. The latency to communicate from one end to the other, however, scales linearly with the number of nodes. If the last node loops back on the first one, then the topology is called a ring.

- *k*-**ary 2-Mesh:** A *k*-ary 2-Mesh is a 2D grid, where each node has links to its four neighbors. The latency of a mesh goes up as a square-root of the number of nodes, and the bandwidth is typically limited by the bisection links.

The topologies described above are used in multi-core processors today. Naturally, they have also found their way into multi-PE DNN accelerators. For instance, Eyeriss [64] and

DNNWeaver [46] use buses, DianNao [65], and ShiDianNao [24] use meshes, SIMBA [61] uses a hierarchical mesh, and TrueNorth [66] uses crossbars and meshes in a hierarchical manner. Note that many more topologies exist in modern systems—trading off latency, bandwidth, and area—which are beyond the scope of this book; we refer interested readers to books on interconnection networks [62, 67].

4.2.2   ROUTING

Routing refers to the specific series of links on the topology that a message traverses to get from its source to its destination. For NoC topologies like Meshes, path diversity between source and destination nodes offers opportunities for choosing one among multiple possible routes to enhance throughput. However, to avoid routing deadlocks [62] (i.e., cyclic dependencies between network packets sitting in buffers) and to provide point-to-point ordering, most commercial NoCs use *deterministic* routing—i.e., a fixed route. The most common routing strategy is XY routing. Discussion on more exotic routing algorithms can be found in books on interconnection networks [62, 67].

4.2.3   FLOW CONTROL

Flow control in NoCs [62] refers to schemes for buffer/link allocation granularity and buffer management.

Buffer and Link Allocation Granularity
In NoCs within CPUs, cache-lines (~64B) are moved around which are much wider than on-chip link widths. Data packets are thus broken down into smaller units called flits (to match the link width). Buffers at each hop can be allocated at the granularity of packets (e.g., store-and-forward and virtual cut-through) or flits (e.g., wormhole). In the context of DNN accelerators, given small operand sizes (8–32 bits), flits are not needed and we assume a store-and-forward style buffer allocation scheme. Detailed discussions of various buffer and link allocation techniques is beyond the scope of this book.

Buffer Management and Control
Traditional general-purpose NoCs connect cores and caches which can have variable delays depending on the application and data availability. Thus, they rely on *back-pressure* between the buffers at connected routers to stall individual packets. Back-pressure is implemented either via ON-OFF/Ready-Valid type signaling between producer and consumer buffers, or via credits. In (On/Off) or (Rdy/Vld) signaling, the consumer holds an ON or Rdy signal whenever it is ready to receive data and turns it off when its buffer becomes full. One challenge is that the depth of the buffers needs to be conservatively sized to account for the delay in transmitting the off-signal—especially if the NoC routers are physically very distant on-chip. In credit flow control, consumers convey the total number of buffers available to the producer before the start

of the traffic flow. Whenever the producer sends a data element, it decrements the credit. After the consumer has finished computing on the data element (or forwarded it further), it sends a credit back to the producer (signaling that the buffer is now free) and the consumer increments the credit. Most DNN accelerators employ one of the flow-control techniques between the L1 buffers and L2 buffers. As discussed previously in subsection 3.2.3, DNN accelerators often perform bulk transfers via credit-based flow-control, paired with a `Rdy/Vld` microprotocol for on/off flow control when consuming the values staged in a buffer.

4.2.4   ROUTER MICROARCHITECTURE

Traditional NoCs routers are multi-ported with input buffers at every port and a crossbar switch. The input buffers are often divided into Virtual Channels (VCs) for various message classes, depending on the coherence protocol. One of the ports connects to the processing core. The router's control path houses a routing unit, and arbitration logic to choose between contending packets for the same output port.

In most DNN accelerators, the router microarchitecture is embedded within the PE itself. PEs employ either partitioned or shared buffers for inputs, weights, and partial sums/outputs, and this is analogous to VCs. The router logic itself is also much leaner. The fixed communication patterns removes the requirement of having complex arbiters—instead PEs simply need to forward data to their neighbors or to the links connecting to the global buffer. A 5×5 crossbar switch like a mesh router is also not needed, and a simple mux to forward data is sufficient.

4.2.5   CHALLENGES WITH TRADITIONAL NOCS

A key challenge with the NoC implementations discussed above is that they are optimized for all-to-all communication—e.g., for cache-coherent multi-processors that deliberately spread data out across different home nodes. These systems are optimized to maximize average-case performance for unknown workloads with arbitrary datasets, in direct contrast to DNN accelerators with statically known neural network layers. Even in an accelerator that switches among several dataflows, NoCs that are provisioned for uniformity can result in simultaneous under-utilization of certain links, and over-saturation of others.

Performance Implication. We implemented and ran the Eyeriss Row-Stationary dataflow traffic through a 8×8 Mesh NoC through all the convolutional layers of VGG16 [3]. Data distribution and collection was performed through a SRAM port connected to the bottom left of the mesh. Figure 4.2 plots the average link utilization. We can observe that most of the mesh links are underutilized. Additionally, there is a large delta of 73% between the highest and lowest utilized link. Both these observations indicate that there is an opportunity for specialized topologies to more properly distribute hardware resources to the bandwidth hotspots.

Area and Power Overhead. The other challenge with traditional NoCs is that they are extremely area and power inefficient compared to the tiny PEs inside DNN accelerators. For instance, a crossbar and Mesh NoC provide high and fair bandwidth, respectively, but require

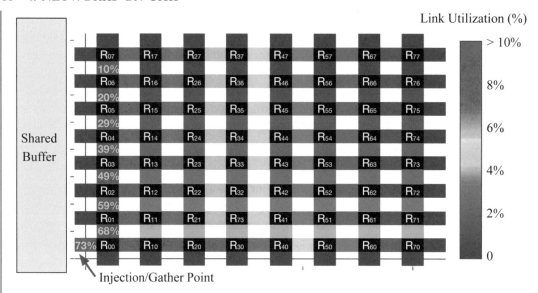

Figure 4.2: Link utilization of 8 × 8 mesh links when it runs 8 × 8 row-stationary [20] dataflow.

497× and 80× more area, while consuming 9× and 35× more power, respectively, than a PE array with 256 Eyeriss [20] PEs as prior analysis has quantified [68]. Buses consume less area and power than the PE array (and are the reason why they have been popular inside accelerators) but lack bandwidth flexibility, requiring them to be carefully tailored at design-time for the dataflow supported by the accelerator.

4.3   SPECIALIZED NOCS FOR DNN ACCELERATORS

Stylized dataflows and communication phases within DNN accelerators leads to a key insight: *efficiently supporting the communication for DNNs within accelerators does not require the same level of generality found in traditional on-chip network routers and topologies seen in many-cores today.* Most of the communication is from the L2 buffer to the PEs to distribute the DNN model and its inputs, and the final outputs need to be communicated back. For a given dataflow, the data movement patterns are deterministic, providing an opportunity to optimize for both performance and energy efficiency by employing specialized NoCs for each communication phase, instead of plugging in a conventional many-core NoC. We discuss design considerations for specialized NoCs in this section.

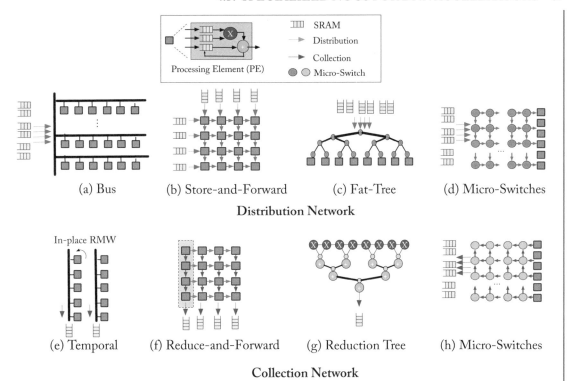

Figure 4.3: NoC implementations for distribution and collection.

4.3.1 TOPOLOGY

Distribution Network

The distribution NoC needs to support two key features: high bandwidth (for performance) to reduce stalls at the PEs waiting for data, and multicast support (to support spatial reuse). The amount of bandwidth and degree of multicasting will depend on the ordering/tiling strategy and parallelism approach respectively within the dataflow. The width of the distribution network links is typically 8–16 bit depending on the input and weight precision.

We describe four classes of solutions for distribution that are used today, shown in Figure 4.3, taking reference from prior accelerator proposals.

Bus. Distribution can be achieved by adding buses to connect the PEs together. For instance, the Eyeriss array uses a hierarchical approach where the PEs along a row are connected by a horizontal ("X Bus") and all X buses tap into a vertical ("Y Bus") that is driven by the global buffer. Data distribution delay depends only on the wire delay to drive the buses, rather than the number of PEs in each row/column. For example, in Eyeriss, the entire distribution takes one-cycle. In a bus-based NoC, operands are inherently broadcast along the entire bus; each

PE snoops and either reads the operand or ignores it, based on the destination ID. Eyeriss adds snoop filters at the entry point of each X-bus to only send the operand if at least one of the PEs on that row is a destination. The bus-based design also decouples the SRAM bandwidth from the physical dimensions of the array. For example, in the Google TPU [25], the number of SRAM banks needed equals the number of rows, while in Eyeriss, it can be independent. Throughput can be enhanced by adding multiple buses. Eyeriss uses four buses for inputs, and one for weights.

Store-and-Forward. Another simple distribution network is to add direct point to point links from SRAM banks to the PEs on the edges of a 2D array, and connect the PEs internally as a grid. This is what the Google TPU [25] uses. Data distribution occurs along a row and/or along a column, depending on the dataflow strategy, in a hop-by-hop manner. Data distribution delay along a row/column is directly proportional to the number of PEs in the row or column (i.e., O(N) if there are N PEs per row or column). However, this delay is pipelined. Both unicast and multicast can be implemented in this network by either forwarding, or storing and forwarding respectively. For instance, in Google's TPU, weights are unicast along the column and kept stationary. Inputs are then multicast horizontally per row.

Fat-Tree. Since distribution occurs from the global buffer to the PEs, a binary-tree can also be used for distribution, with the global buffer at the root and the PEs at the leaves. A binary tree can support arbitrary unicast, multicast, and full broadcast by setting switches appropriately at each level, thereby providing high flexibility for mapping different dataflows. Higher bandwidth can be achieved by employing a fat-tree. In theory, distribution of each unique operand (unicast,multicast, or broadcast) through a tree with N leaves takes $O(\log_2 N)$ cycles. However, if the data distribution path is preset (based on the dataflow), the intermediate switches do not have to latch the values and the number of cycles to traverse the tree can be set by the wire delay [68], rather than the number of levels.

Custom Micro-Switches. Instead of using traditional NoC topologies, one approach for distribution has been to build custom topologies using simpler grids of tiny switches [58, 68, 69]. We call these micro-switches, to distinguish them from switches/routers used in traditional NoCs. The microswitches are built using a few muxes, and do not require expensive buffers and arbitration logic like conventional routers. For instance, Eyeriss v2 [69] uses a Hierarchical 2D Mesh of such switches for data distribution across multipler clusters of PEs. There is an all-to-all network within each cluster. The data distribution occurs in a hybrid manner with hop by hop unicast/multicast/brodcast over the Mesh, followed by a single-cycle unicast/multicast/broadcast to any PE within the cluster. A fat-tree based topology build using micro-switches is used within the MAERI [58] accelerator, with two optimizations over a conventional fat-tree. First, the bandwidth at the upper levels of the tree can either be 2× (like a fat-tree) or the same as that at the lower levels. This essentially allows the root bandwidth to be set independent of the number of leaves. It is called a "Chubby Tree." Second, MAERI's chubby

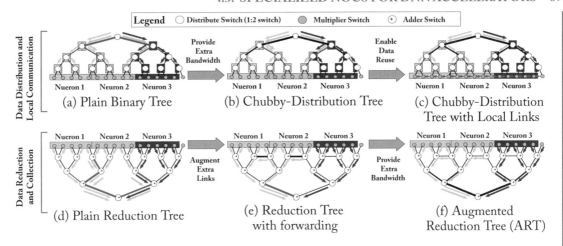

Figure 4.4: Distribution and collection networks in MAERI [58].

tree also adds store-and-forward links at the leaves. MAERI's distribution network is shown in Figure 4.4a–c.

Collection Network

The collection NoC needs to support two key features: high bandwidth (for performance) to deliver all outputs to the global buffer without stalls, and reduction support (to support spatial reuse). The amount of bandwidth and degree of reduction will depend on the ordering/tiling strategy and parallelism approach respectively within the dataflow. The width of the collection network links is typically the same as, or wider than the distribution network, since it holds accumulated values that may need higher precision.

We describe four classes of solutions for collection used today, shown in Figure 4.3, taking reference from various accelerator proposals.

Temporal using RMW Buffer. The simplest reduction approach is temporal in-place reduction via an output stationary dataflow. Reduction time is O(N) for a N-sized dot-product. This approach does not require the support for spatial reduction, and the collection network simply needs to read all the outputs out from each PE and deliver to the global buffer, which can be done via any NoC topology (e.g., a collection bus in ShiDianNao [24]).

Reduce-and-Forward. Spatial reduction of partial sums can be performed via a linear reduce-and-forward approach. Reduction time is O(N) for a N-sized dot-product. This approach is used by the Google TPU and Eyeriss for reduction across each column. The outputs from each column can be collected by the global buffer (if complete) or sent to RMW buffers for further temporal accumulation [25].

Reduction Tree. Spatial reduction of partial sums can also be performed by connecting the adders via a tree. Tree-based reduction is employed in Microsoft's Brainwave [70] and NVIDIA's NVDLA [26], and requires O($\log_2$N) time for a N-sized dot-product.

Custom Micro-Switches. Custom topologies for the collection network can also be connected via micro-switches, similar to distribution. For example, MAERI [58] uses a novel reduction tree topology called Augmented Reduction Tree (ART) that adds forwarding links between sibling adders that do not share a parent. These links are used to forward partial sums to different branches of the adder tree to allow reductions that are non powers of two in size. The root of the tree generates outputs that need to be written back to the global buffer. Multiple outputs can be collected together via fat-links. MAERI's collection network is shown in Figure 4.4d–f. Similarly, the Eyeriss v2 hierarchical mesh enables collection of outputs from individual groups and forward them to the global buffer.

Topology Selection and Channel Width

An accelerator designer can choose any pair of distribution and collection topologies described above for their accelerator, as long as it can meet the connectivity requirements from the expected mapping strategies. It is also possible to share the same network links for both input and weight distribution as virtual channels, or employ separate physical networks for each datatype, depending on the bandwidth requirement. In inference accelerators, typically the input and weight operands are 8–16b. The distribution links are sized to be the same width as the operands. This is unlike NoCs for multicores where 64B cache lines often need to be divided across multiple flits and transmitted over narrow links. Most DNN accelerators use a separate collection NoC, but it is certainly possible to share the same links for both distribution and collection. For example, the vertical links in a TPU are used for weight distribution, and for partial-sum reduction. One caveat, however, is that the outputs have higher precision than inputs or weights, and so the collection links need to be wider.

Flexible Aspect Ratios

Depending on the physical NoC topology employed for distribution and collection, it can be morphed into different logical topologies to provide flexibility during mapping. For example, consider a 4 × 4 PE array shown in Figure 4.5a. Suppose the PEs are connected via a rigid store-and-forward network for distribution (Figure 4.3b) and a reduce-and-forward network for collection (Figure 4.3f). For simplicity, we consider a GEMM mapping over this array.

Figure 4.5b demonstrates how a mismatch in dimensions between the matrices and the physical array leads to under-utilization of the compute elements, and requires two passes to finish the computation. This inefficiency arises due to two reasons. First, the store-and-forward distribution NoC multicasts each row of the blue matrix across one row of the physical array. Second, the reduce-and-forward NoC reduces all partial sums across the column into one output. In contrast, as shown in Figure 4.5c, leveraging a flexible distribution NoC that can multicast

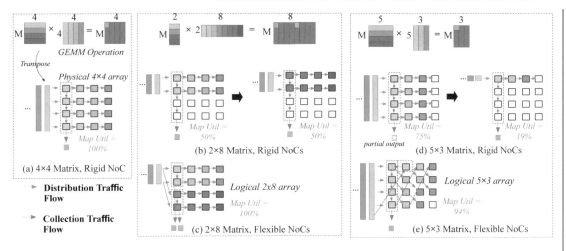

Figure 4.5: Distribution and collection NoC support for enabling flexible aspect ratios.

the same data across multiple rows of the physical array (say a bus, fat-tree, or one built using custom micro-switches) and a flexible collection NoC (using a custom topology) that allows tapping outputs from PEs at the middle of the array can help morph the physical 4 × 4 array into a logical 2 × 8 array. This enhances utilization and finishes the computation within one pass.

Figure 4.5d, e demonstrates a similar case for a fixed 4 × 4 array morphing into a 5 × 3 array, but requires even more flexibility from the distribution and collection NoCs to support the required traffic flow.

4.3.2   ROUTING

The NoC topologies employed within DNN accelerators naturally lead to a deterministic routing approach, both during distribution and collection.

4.3.3   FLOW CONTROL

Latency-Insensitive and Latency Sensitive Control

Recall from subsection 4.2.3 that stalls within the NoC are implemented via backpressure signals. This is known as *latency-insensitive* paradigm. This is the norm in multi-core NoCs. It is also the buffer control mechanism in many DNN accelerators like Eyeriss [20]. To put it simply, latency-insensitive accelerator implementations allow each PE to stall independently, which might occur either due to some PEs taking longer than others to produce data, or the NoC not having sufficient bandwidth to keep all PEs active.

On the other hand, the static knowledge of the communication flow once the mapping is determined can also allow DNN accelerators to operate in a *latency-sensitive* paradigm. This is leveraged in the Google TPU [25], which we discuss later in Chapter 5, and is called a **Systolic Array**. Systolic arrays operate by forwarding operands every cycle between the PEs, without waiting for flow-control signals. Each PE is guaranteed to compute a partial sum within a cycle, and there is no dynamism within the PE. To implement such stall-free accelerator arrays, both the distribution and collection NoCs need to have sufficient bandwidth for the dataflow strategy. In systolic arrays, if the global buffer needs to stall for DRAM, then the *entire accelerator array* stalls.

A key implication of the flow-control mechanism is centralized vs. distributed nature for the accelerator control. Latency-insensitive accelerator implementations allow fully distributed control, while latency-sensitive accelerators need a central controller to reserve buffers at the end points, and orchestrate the communication.

Link Reservation Flexibility

As we discuss later in Section 4.4, the dataflow determines the bandwidth and degree of spatial multicast/reduction within a DNN accelerator. This in turn introduces the question of the granularity at which the NoC resources (links and latches) can be reserved. We classify the reservation choices into three categories: Dedicated, Shared+Configurable, and Shared+Arbitrated.

Dedicated is the mode of operation where the traffic-flow over each NoC link is determined at *design-time* and cannot be changed during the operation. For example, the weight stationary dataflow approach over the TPU is an example of such a reservation, since there is a one-to-one correspondence between which SRAM bank feeds which distribution link.

Shared + Configured is the mode of operation where the dataflow (i.e., the loop order and/or partitioning approach) can be changed at *compile/map-time* and the NoC is configured accordingly to handle different traffic flows. This is used by Eyeriss and MAERI. For instance, in Eyeriss, the effective aspect ratio of the array can be configured when mapping the DNN (as shown in Figure 4.5) by updating the snoop filter configuration within the NoC. In MAERI, the size of virtual neurons (group of PEs computing a dot-product) can be configured depending on the DNN model and partitioning strategy.

Shared + Arbitrated is the mode of operation where the dataflow can be changed at a fine-granularity at *runtime*, requiring a fully arbitrated NoC. This is typically not common in DNN accelerators, as described so far. However, this approach might make sense if the accelerator supports activation sparsity, and the data delivery approach and corresponding traffic directly depends on the value of the input.

The right level of flexibility within a DNN accelerator's NoC and the corresponding approach for reserving the links at design/compile/runtime is an open area of research currently [68, 69].

4.4   LEVERAGING REUSE VIA THE NOC

As we have discussed in previous chapters, memory accesses (reads and writes) are one of the major energy consuming operations in DNN accelerators. Exploiting data reuse is crucial for obtaining high energy efficiency without compromising performance. Recall from Chapter 2 that algorithmic reuse gets exposed based on the dataflow strategy being employed. The amount of temporal reuse and spatial reuse that the accelerator can exploit depends on the dataflow strategy. This in turn has a direct implication on the traffic flow within the NoC of the accelerator. We discuss this connection next.

4.4.1   IMPLICATIONS OF TEMPORAL REUSE

As discussed in Chapter 2, temporal reuse refers to opportunities for reusing operands within a buffer. To exploit temporal reuse, accelerators often employ a memory hierarchy that consists of a shared global buffer, which we refer to as L2 or global buffer, and private buffers for each PE, which we refer to as L1 or local buffer, as Chapter 3 discussed.

Temporal Reuse during Inputs and Weights Distribution. Temporal reuse of operands within a PE determines the bandwidth requirements of the network *from the global buffer* during distribution, since new operands need to be fetched once the PE finishes reusing the ones buffered within the PE. High temporal reuse can, in principle, reduce the bandwidth requirement. However, since there may be hundreds of PEs in the accelerator, the aggregate bandwidth requirements can remain quite high. For example, suppose each PE reuses a hypothetical L1 tile for 420 cycles and needs 21 new 16-bit operands for every new tile. This makes the bandwidth requirement from the PE $(21/420) \times 16 = 0.8$ bits per cycle. In a 100 PE system, however, this translates to 80b reads from the L1 buffer every cycle. Any congestion in the network could easily make data delivery the rate-limiting step.

Temporal Reuse during Output Collection. During collection, there can be two forms of communication: between PEs (for local partial sum accumulation) and from PEs to the global buffer (for reduction of multiple partial sums). The bandwidth requirement of the collection network thus depends on the number of outputs that need to be written *to the global buffer*. A pure output stationary dataflow, that relies on temporal reuse via a read-modify-write buffer within each PE for in-place reduction, can eliminate inter-PE reduction completely, eliminating the need to provision for any inter-PE bandwidth within the collection NoC. However, in such a case, the bandwidth requirement at the write ports of the L2 during the collection phase could be very high as each PE generates a unique output. Moreover, a pure output stationary dataflow will also need a unique weight and input every cycle, increasing the bandwidth requirement of the distribution NoC. In contrast, if spatial reuse is leveraged for reducing the outputs, as we talk about next, the bandwidth requirements during distribution and collection may go down.

4.4.2 IMPLICATIONS OF SPATIAL REUSE

As discussed in Chapter 2, spatial reuse refers to opportunities for reusing operands across the PEs. Spatial reuse fundamentally requires devoting bandwidth resources to support it, otherwise such a mapping strategy becomes invalid.

Spatial Reuse during Input and Weight Distribution. Spatial reuse of input operands (i.e., activations and/or weights) results in one-to-many traffic (multicast) from the shared L2 buffer to the PEs. It can be exploited via explicit support for multicast (or broadcast) via the NoC, enabling delivery of the operands read from the global buffer to multiple or all the PEs. This dramatically reduces the number of operand fetches from the global buffer. Distributing the operands spatially over wires, rather than temporally reading them over and over again from a large global SRAM buffer can often lead to reduced energy consumption (though this depends on the size of the buffer vs. the length of the interconnects). In addition, multicasts can also enhance performance, since data delivery time can be reduced by a factor of N (where N is the number of PEs that required the same operand). Multicast is thus a critical bandwidth reduction component, as it amortizes the network delivery, the L2 address calculation, and the L2 access energy. Multicast support can be implemented in the NoC as a store-and-forward over the PEs (e.g., Google TPU [25]), or via explicit fanout topologies such as trees (e.g., MAERI [58]), as discussed earlier in Section 4.3.

Spatial Reuse during Output Collection. Another spatial reuse opportunity, namely many-to-one or reduction, occurs during partial sum accumulation. Partial sums are individual multiplication results of weight and input activation as described in Figure 1.5, and they need to be accumulated to generate an output activation value. Unless an output stationary dataflow is employed, partial sums for the same output are generated by different PEs. One approach that SIMD lanes in GPUs employ, is to send all the partial sums to the global buffer and have a separate accumulation phase. Instead, NoCs in DNN accelerators reduce the partial sums via explicit support such as a linear forwarding and addition through the PEs (e.g., Google TPU [25]), or a spatial adder tree (e.g., NVDLA [26] and MAERI [58]), as discussed earlier in Section 4.3. Providing such special connectivity between the PEs can save buffer writes, and, in case of an adder tree, reduce the reduction time from $O(N)$ to $O(\log_2 N)$.

4.5 TYING IT TOGETHER: FROM DATAFLOW TO TRAFFIC FLOW

We go through a detailed walkthrough example tying in the concepts discussed in this chapter. Figure 4.6a, b show two example mappings of CONV1D operation introduced in Section 2.2 on a three-PE accelerator, and the corresponding traffic flows for each mapping. We can observe that traffic flows from both mappings consist of the distribution and collection communication phases we discussed in Section 4.1. However, the fine-grained data movement schedule within each phase and amount of data movement differs based on the mapping.

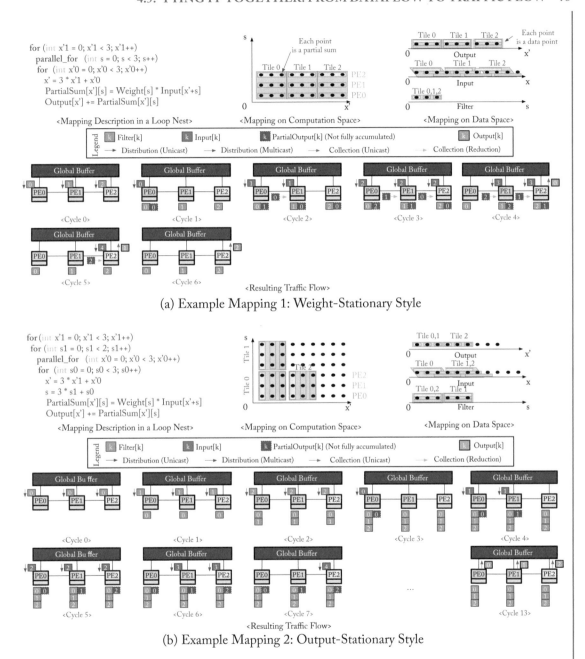

(a) Example Mapping 1: Weight-Stationary Style

(b) Example Mapping 2: Output-Stationary Style

Figure 4.6: An example mapping of CONV1D operation on a three-PE accelerator and corresponding traffic flow within tile 0 of each mapping.

In Figure 4.6a, a weight-stationary style mapping is used. The filter weights are distributed via unicasts at the beginning and remain stationary. The sliding window behavior of convolutions shows up as a spatial reuse opportunity in this mapping and can be leveraged via multicast of the input activations (except those at the boundaries). This can be implemented via store-and-forward between the PEs, or a fanout from the global buffer (as Figure 4.6a shows), discussed earlier in Section 4.3. Partial sums generated by the individual PEs are reduced spatially, which can be implemented via a reduce-and-forward (as Figure 4.6a shows) or an adder-tree, as discussed in Section 4.3. The final outputs are collected from PE3 cycle-by-cycle. The global buffer thus needs a write bandwidth of one element per cycle, as is visible Cycle 4 onward in the example.

In Figure 4.6b, an output-stationary style mapping is used. There is spatial reuse opportunity for both inputs and weights during distribution across the PEs. The weights are distributed via a broadcast to all PEs, while the input activations are multicast from the global buffer. There is no spatial reuse during collection in this mapping, unlike the previous example. Each PE performs a temporal reduction and generates a unique output. The outputs from all PEs are collected at the end of the computation phase. The write bandwidth requirement into the global buffer is higher (equal to the number of PEs) and bursty in this mapping, compared to the previous mapping.

4.6   SUMMARY

In this chapter, we discussed the design approach for moving operands and outputs on-the chip between the memory and processing elements using on-chip networks. In particular, we identified the relationship between dataflow and communication, and opportunities for optimizing the data distribution and collection on-chip networks. By combining together datapaths, buffers, distribution, and collection networks we now can create complete neural network accelerators. The specific arrangement of these pieces depends on the desired dataflow and mapping, as we discuss in the next chapter.

CHAPTER 5

Putting it Together: Architecting a DNN Accelerator

In the previous chapters we have discussed the various crucial parts of building a neural network accelerator one by one. In this chapter we zoom out and conceptualize how all the pieces bind together and work as a synergistic system optimized for various design goals.

Accelerators, as the name suggests, are primarily designed for performance. In the case of DNN acceleration, as we have discussed so far, performance is extracted by exploiting the parallelism offered by the algorithm. Naturally, therefore, DNN accelerators consist of a collection of MAC units which run in parallel. Increasing the performance of the accelerator translates to having more and more of these MAC units in the design. Managing a large parallel design however is not trivial. As we discussed in the previous chapters, ensuring that all the MAC units are fed with useful compute requires careful design of the memory and interconnect system. Moreover, eliminating wasteful data fetches via reuse to extract efficiency from a system depends on the dataflow and mapping.

In the following sections we first discuss the flow employed while designing a DNN accelerator. We also discuss high-level design decisions one needs to make when architecting for specific use cases. In the end we conclude the chapter by performing case studies of a few well known accelerators. We reason about the implementation choices made in these proposals and ponder about the implications in term of performance and efficiency.

5.1 DESIGN FLOW

We break down the design-flow for a DNN accelerator design into several steps, as highlighted in Figure 5.1. We discuss these below.

5.1.1 TARGET SPECS AND CONSTRAINTS

When building a custom accelerator, there are two categories of *specs* which dictate the design choices—workload and performance—which need to be satisfied within a set of *constraints*. Table 5.1 lists some examples for performance specs and design constraints, at the time of writing this book.

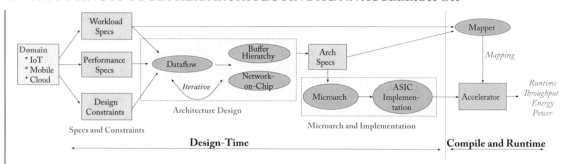

Figure 5.1: DNN accelerator design flow.

Table 5.1: Estimates of technical specifications and constraints for accelerators suited for various classes of use cases

Specs and Constraints	IoT Class	Mobile Class	Data-center Class I	Data-center Class II
Operations	Inference	Inference	Inference	Training
Peak throughput	100 GOps	1–10 TOps	~100 TOps	~100 TFlops
Target inf. latency	10 ms	15 ms	100–250 ms	100–250 ms
Power	<= 1 W	1–10 W	100 W	300 W
Area	<10 mm^2	1–10 mm^2	~100 mm^2	~100 mm^2
Memory BW	None	DRAM	HBM	HBM

Workload Specs

Depending on the deployment use case, the design of the accelerator can be optimized to run only one specific DNN (e.g., an accelerator within a drone running YoloTiny [7] for object detection and tracking), a handful of specific DNNs (e.g., an accelerator on a AR/VR device [71] running Resnet for image classification and UNet for image segmentation), a specific class of DNNs (e.g., an accelerator on a mobile phone running various vision algorithms using dense CNNs), or be as general as possible for current and emerging DNNs (e.g., an accelerator in the datacenter running myriad vision, speech, language and recommendation DNNs).

The workload specs determine the following: (i) primitive compute approach—convolutions vs. matrix-matrix multiplication vs. vector-matrix multiplications within the accelerator; (ii) precision of the compute units (depending on the accuracy of the trained DNNs this accelerator is expected to run); (iii) which layers to support efficiently; and (iv) special functional units.

For instance, suppose we wish to design an IoT class accelerator that only runs YoloTiny. We can go with a convolution compute model with 8-bit fixed point MAC units. Alternately, if we wish to design a datacenter class accelerator that runs both inference and training, we may have to go with the GEMM compute model with floating point MAC units (like the Google TPU v3).

A key challenge today in deciding the right workloads to optimize the accelerator for, especially for datacenter scale accelerators, is that the workloads of interest are themselves changing at a rapid scale and becoming evermore complex. This is because, at a high level, improving the accuracy of a neural network generally translates to adding more parameters or weights. Therefore, a better performing network requires more computations. For example, proposals in language models have hundreds of millions [12] to several billion operands [72] which demand an extremely high performance for fast execution.

Performance Specs

Depending on the end application that is using the DNN model, the accelerator would typically come with performance specs—such as peak throughput or maximum latency. For instance, a certain frame rate of inferences for a given image resolution may determine the target throughput, or a use-case within real-time systems such as a self-driving cars may dictate the expected latency.

The throughput specs directly determine the number of MAC units we need. The latency specs affect the dataflow strategy, and the bandwidth and memory requirement to remove stalls.

For example, let us consider a target peak throughput of 512 GOPs at 0.5 GHz. This translates to 512×10^9 Ops per second at 0.5×10^9 cycles per second translates to 10^3 Ops per cycle or in other words 10^3 MAC units. For the sake of convenience we can allocate 32×32 (i.e., 1024) MAC units in this accelerator.

Similar considerations come in for latency specs. Suppose we have an end-to-end latency requirement of 10 ms for running a small model (e.g., MobileNet_v2) on a mobile accelerator; we need to put 64 MB on-chip SRAM to fit MobileNet_v2's weights to eliminate memory fetches (except at the beginning when loading the model) and focus on efficiently streaming in input data from sensors to the accelerator. On the other hand, suppose we are designing a datacenter accelerator. DNN inferences are only one of the several operations being carried out in a single query. In recent years, web applications are composed as a collection of micro services [73]. In general, use of micro services introduces extra network traversals and library invocations as compared to a monolithic web application. This exacerbates the latency bound for a single operation (e.g,. DNN inference) and in demands stricter SLO (service level objective). Furthermore, the end-to-end runtime also includes several tasks like data transfer, IO operations, and several services, making the inference bound even stricter.

Design Constraints

Depending on the deployment use-case (IoT vs. mobile vs. cloud), the accelerator will need to be designed with certain constraints in mind—such as (i) area, power, energy budget; (ii) available memory capacity and memory bandwidth; and (iii) system level constraints such as the interaction approach with the rest of the system (e.g., separate memory for accelerator vs. shared cache with host, decoupled vs. coupled access-execute, and so on). Table 5.1 lists some example constraints for modern accelerators.

The area-power-energy consumed by the accelerator depends on the compute elements (number and precision), as well as the data orchestration mechanisms (dataflow, memory hierarchy, and interconnect). We go over techniques to determine these overheads in Chapter 6. The art of accelerator design is in achieving the specs while meeting these constraints. Thus, the overall efficiency of the accelerator often gets compared via metrics such as TOPS/W or TOPS/mm².

5.1.2   ARCHITECTURE DESIGN

Building DNN accelerators translates to a design with a large number of MAC units based on the specs. In order to keep all the MAC units busy, several factors are needed to be considered. First, the dataflow mechanism should ensure that relevant computation is being mapped to as many MAC units as possible at all time. Second, appropriate amounts of memory should be allocated at the PEs such that the tile of operands required for running the mapped computations are available. Third, all the scheduled compute units should be able to receive required operands in their buffers at the right time to perform relevant work. Stalling compute units for data undermines the benefit obtained by reduction of compute time while incurring higher costs due to wasted power. Fourth, is the cost of scaling. Increasing the computing power generally translates to the addition of more hardware, which not only comprises of the MAC units themselves, but also the additional control logic, memories, interconnection networks and so on. Each added hardware unit translates to area and power consumption, which in turn drive up the cost of manufacturing and the cost of operation. We go over these next, following the flow shown in Figure 5.1.

Dataflow

The primary consideration is the dataflow of the accelerator, to ensure that relevant computation is being mapped to as many MAC units as possible at all time. Recall that the dataflow depends on the tiling strategy, loop order, and parallelization strategy.

The tiling strategy sets the number of buffering levels. The loop-order determines the rate at which different dimensions of each tensor need to be fetched from and written to the next level of the buffer hierarchy. The outermost loops within each tiling level determine the *stationary* tensor—the one that is fetched at the slowest rate. For instance, given that weight matrices have extremely high memory footprints and offer ample reuse, especially if batching is used, employing a weight stationary dataflow is a highly popular choice in both NVDLA and TPU [25, 26], as we discuss later in this chapter. The parallelization strategy determines which loops get mapped across the spatial dimensions of the accelerator. This is highly workload and layer dependent. For image classification CNNs (e.g., VGG16, ResNet), the network starts with high-resolution images with shallow channels, but the number of channels increases as we go deeper into the network. Moreover, the number of weight elements also increase and surpass the number of input activation elements as we go deeper into the network. This suggests that there is a lot of parallelization opportunity across input and output channels. This is exploited by NVDLA [26], as we discuss later in this chapter.

It is possible to design accelerators with a single dataflow [25, 26], or multiple dataflows [42, 58]. In the latter case, the buffering and NoC implementations need to be accordingly architected to be able to handle all possible dataflow strategies.

Buffer Sizes

Given the dataflow, we need to determine the buffer sizes to keep the datapath fully utilized. These buffer sizes can be determined by reuse-aware cost models that are discussed in Chapter 6. For instance, a dataflow that keeps the filter row stationary (e.g., Eyeriss) but does not have enough buffers within a PE to store the row would be a legal but a highly inefficient implementation of the dataflow since elements of the row would need to keep getting fetched. For the stationary tensor, we can determine the coupled dimensions (discussed earlier in Chapter 2) for that tensor and compute the size of the buffer to hold the tiles that will remain stationary.

Next, given the size of the buffer for the stationary tiles (say weight), the sizes of buffers for the other tensors (input and output) would need to be determined to remove/reduce spills to the next level of the memory hierarchy.

Network-on-Chip Bandwidth and Latency

Given the dataflow (stationary operands and partioning strategy) and the buffer sizes at each PE, the amount of data that needs to flow in and out of the array can be computed independently for the weights, inputs, and outputs. Spatial reuse opportunities provided by the dataflow can help reduce the bandwidth requirement from the L2 buffers since the same data element can be multicast within the NoC. Similarly, spatial reduction within the collection NoC can reduce the number of unique outputs that need to be written back to the L2 buffer. The dataflow thus helps determine the overall *bandwidth requirement* from the distribution NoC and collection NoC to deliver data at a target rate to maintain the overall throughput specs. The amount of temporal data reuse at the PEs can help compute the *latency budget* for data distribution and collection, by overlapping computation and communication.

The overall bandwidth requirement and latency budget can help determine the right NoC topologies to use for data distribution and collection.

Hierarchies and Scalability

The dataflow, buffer, and NoC analysis needs to be repeated for every level of the tiling hierarchy in the accelerator system. Accelerators for IoT and mobile class accelerators with hundreds of

PEs typically use only 1–2 levels of the memory hierarchy. However, datacenter scale accelerators with thousands of PEs could either be scaled-up using a single level of hierarchy (e.g., each 128×128 array in TPU v3), or scaled-out by connecting multiple arrays together [61, 70, 74–76]. Naturally, scaling the compute, memory, and interconnect in the right proportions, while meeting the performance specs and design constraints is a non-trivial task.

5.1.3   MICROARCHITECTURE AND IMPLEMENTATION

Given the architectural specs discussed above, next the individual components of the accelerator need to be implemented. The microarchitectural choices are driven by two factors: the desired functionality and the area-power-energy constraints.

The PE datapath is constructed by determining the number of MAC units within each PE, their connectivity (individual SIMD lanes vs. a dot-product), and the precision. Typically, the precision of the weights and inputs would be lower (e.g., 8-bit) while the precision of the output would be 16–32 bit.

Next, the buffers can be implemented as simple FIFOs, scratchpads, or via emerging paradigms like buffets [32], as discussed in Chapter 3. The choice of the specific microarchitecture is driven by the dataflow, as it dictates the functionality of the buffers (e.g., FIFO access vs. random access). Moreover, the buffers could be private per PE, or shared by a group of PEs, depending on how the dataflow is designed. For the same functionality, this choice is further determined by the area or power of the underlying circuit implementation, which often depends on the size of the buffers needed (e.g., a flip flop is probably the best for a L1 buffer of size 1, while larger L2 buffers need SRAM scratchpads).

Finally, the NoCs can be implemented. As described in Chapter 4, the distribution and collection networks may use myriad topologies such as buses, store/reduce-and-forward links, or trees. Similar to buffers, the implementation choice depends on both the dataflow (which determines desired functionality) and the underlying circuit (which determines area and power cost)

Given the final microarchitecture, the various components are implemented in RTL, and sent through an ASIC flow (synthesis, place-and-route) to determine overall area and power. If the area-power constraints are not met, the microarchitecural choices can be re-evaluated. If alternate microarchitectures do not meet the area/power budget, the designer would have to go back one more step and re-architect the accelerator.

5.1.4   MAPPING

Once the final accelerator is taped out, the compile/runtime flow requires mapping the workload over the accelerator, as shown in Figure 5.1. A mapper takes the workload specs and the architectural specs as an input, tiles the workload, and schedules each tile spatially and temporally, honoring the dataflow strategy of the accelerator. The efficiency of the accelerator microarchitecture and the mapping together determines overall runtime, throughput, energy, and power

Table 5.2: CONV2D layers we use in walk-through design examples

Layer	K	C	Y	X	R	S	Note
CONV5_2	512	512	7	7	3	3	Optimization Target
CONV2_2	64	64	56	56	3	3	Alternative Workload

consumption when running the desired workload on the accelerator. For accelerators with a fixed dataflow, the role of the mapper is to determine optimal tile sizes at each buffering level for the given workload, such that the tiles fit across the accelerator's memory hierarchy. However, if the accelerator is more flexible/programmable and supports multiple dataflows, the role of the mapper becomes even more crucial as it needs to pick both the right dataflow and the right tile sizes for that dataflow. We discuss more details of the mapping step in Chapter 6.

5.2   EXAMPLE DESIGN WALK THROUGH

In this section we will put the steps discussed in previous section into work and illustrate the design-flow of two example accelerators—one data center class and one mobile class. For these examples, we will walk the readers through a simple hand-designed flow. In Chapter 6, we will go over mechanisms for automating this flow and searching through the design-space for optimal configurations.

Step 0: *Workload Specs.* For both accelerators, for simplicity we pick the CONV5_2 of ResNet50 [5] as the target workload. After we finish designing an accelerator, we run the design on another layer, CONV2_2 of Resnet50, to observe how the performance changes with another layer. Table 5.2 describes the layer dimensions of those two layers.

5.2.1   DATA CENTER CLASS EXAMPLE

For our first example we choose a data center class target. For the sake of simplicity we will try to design a Google TPU-like system. The final design is summarized in Figure 5.2. We walk through the design-flow next.

Step 1: *Performance Specs.* A data center class accelerator needs to allow for high throughput as well and low latency. For this reason, such accelerators need to maximize for parallelism by employing a large number of MAC units. Let's assume for our case we need about 32 TOPS @1 GHz, with each MAC operation being counted as two operations. This therefore translates to a system with 16 K MACs (16384 to keep the number a power of 2) similar to the performance of Google TPU v2 system at 1 GHz.

Step 2: *Dataflow and Mapping.* Given a datacenter class accelerator that needs to run a variety of DNN architectures (not just CNNs), we go for GEMM as the computation being accelerated.

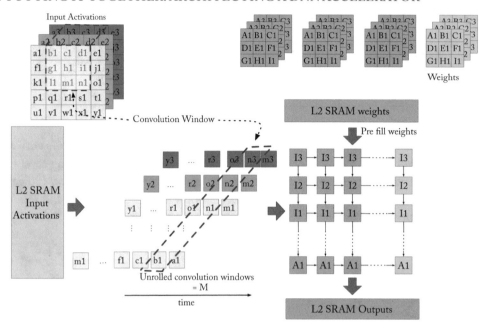

(a) Data Center Class Example Accelerator

Frequency	1 GHx
Performance	32 TOPs
Number of MACs	16384
Dataflow (order)	Weight Stationary
Dataflow (partition)	(K-N)*
L1 Size	49 KB
L2 Size	23.6 MB
Distribution NoC	Store and Forward (BW: 128 GBps)
Collection NoC	Reduce and Forward (BW: 128 GBps)

(b) Design parameters for Data Center Class Example Accelerator

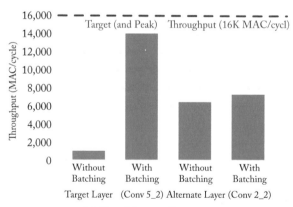

(c) Resulting performance

Figure 5.2: Schematic showing (a) the data orchestration in an example data-center class accelerator—the mapping in the temporal and spatial dimension are determined by the operand matrix dimensions in the GEMM; for our target layer we get M = 25, N = 512, and K = $3 \times 3 \times 512 = 4608$; (b) the specifications for designing the example data-center class accelerator (* *K-N refers to the GEMM dimensions mapped on the spatial dimensions of the array; where K is the number of elements in a convolution window and N is the number of filters*); and (c) chart depicting the overall throughput achieved for the target and alternate layers, with and without batching.

Recall from Figure 1.2 in Chapter 1 that FC layers can naturally be represented as GEMMs, while converting a convolution to a GEMM was discussed in Section 2.7. Choosing the dataflow is a non-trivial exercise which we have discussed in Chapter 2. We go with the dataflow strategy employed by the TPU.

Parallelized Dimensions. We partition the filters spatially across the columns of the array. This is the N dimension of the GEMM operation. The inputs are tiled into convolution windows and spatially partitioned across the rows of the array. This is the K dimension of the GEMM operation. Figure 5.2a shows this partitioning. Input activations get replicated across the convolution windows as the GEMM implementation of CONV2D loses reuse across the sliding window.

Tile Scheduling (Loop Order). Each weight filter is unrolled in the RSC order, and fed into the PEs from the columns of the array, as Figure 5.2a shows. We implement a weight stationary dataflow by keeping tiles of the $N \times K$ (i.e., weight) matrix stationary within the PEs, while the convolution windows (M dimension of the GEMM) are streamed in temporally via the rows of the array. The loop-order is thus N-K-M or K-N-M.

Tile Sizing. The tile of data kept stationary depends on the size of the array. In our design, the tile size for each weight filter is *row* (the number of rows of the array, i.e., 128), and the tile size of the filters is *col* (the number of columns, i.e., 128). These *row* $\times$ *col* weight remain stationary. Similarly, input tiles of size *row* are streamed in to the array to be multiplied with the weights in each column; each column produces a partial (or full) output after accumulating all the partial sums spatially.

Step 3: *Buffer Allocation.* In this step we will determine the sizes of the L1 buffers within the PEs and outside the array (L2).

L1 Scratchpad Buffer. In the datacenter accelerator with 16 K PEs, it makes sense to keep each of the PEs as small as possible. This helps with the scalability of the design as it is flexible to layout and connect simpler compute units, which is important for building a large accelerator. With this design philosophy, we choose a PE design which is capable of doing one MAC per cycle and minimum buffering: a one word wide register to store one element of the stationary weight operand, a one word wide register to store the streaming input operand, and a one word wide register for storing the generated partial sum. It has been shown in the works by the machine learning community that for inference workloads, a word size of 8 bit is sufficient to operate with without sacrificing reasonable accuracy [77]. Thus, the L1 buffer in each PE contributes to 3 bytes of capacity. The whole array therefore has 49 KB ($3 \times 16384 = 49152$ bytes) of storage.

L2 Scratchpad Buffer. Deciding the L2 buffer size in our case depends on the dataflow employed.

- *Filters.* As the filter data remains in the L1 buffers for a substantial amount of time, we need a small L2 buffer to stage data for the next swap. In our case that would be 16 KB for weights (since 16 K MACs each holds one byte of the weight filter).

- *Input Activations*. For inputs, however, which are streamed in every cycle, we provision enough L2 buffering to store the entire input matrix. Moreover, as the layer is executed as a GEMM operation, there will be elements of the input which will be replicated across rows of the input operand matrix which correspond to the different rows in the GEMM matrix, as shown earlier in Figure 2.13. In other words, the size of the input matrix would be $K \times M$ (i.e., *elements per convolution window × pixels per output channel*). Thus, taking these factors into account, the size of the input matrix for the target layer (CONV_5_5 of Resnet-50) comes out to be 115 KB. Datacenter accelerators also often employ batching to improve the throughput. However, batching requires extra L2 capacity. We provision for a batch size of 100 and thus allocate 115 KB ×100 = 11.5 MB for input activations.

- *Outputs*. Given the weight stationary mapping of computations in this example, each column of the array is responsible for generating one output pixel. We choose the L2 size to hold all the output pixels (which becomes the inputs for the next layer). For every pass, the maximum number of pixels or partial sums generated therefore is equal to the product of the number of pixels per output channel and number of columns of the array. For the target CONV_5_5 layer in Resnet-50, this is equal to 3.2 KB. When considering batching, we need to provision for the batch size worth of outputs generated. As we chose a maximum batch size of 100 the amount of output L2 allocated is equal to 320 KB.

This brings the total size of L2 buffer to be 11.8 MB, and to 23.6 MB when considering double buffered access. Given that we are designing for a data center class processor, it is wise to over-provision the buffer for workloads with various arithmetic intensities and operand sizes.

Step 4: *Network-on-Chip.* For the target dataflow, the distribution phase requires the stationary filters to be fed into the array, and each convolution window (fed across the rows) from the streaming matrix to be multicast to all the filters. We implement this using a store-and-forward network across the columns (for the stationary matrix) and across the rows (for the streaming matrix) to keep the wire area-power overhead low for the large design. During the collection phase, the partial sums generated by each PE need to be accumulated spatially across the rows off each column. The spatial accumulation is implemented via a reduce-and-forward linear network to keep the wire area-power overhead low for the large design. The bottom PE in each column generates the full (or partial) output activation. If the generated outputs are partial (if the dimensions of the weight matrix exceed those of the array), further in-place accumulation is done at the L2 to generate the full output.

The distribution and collection can be performed in parallel by carefully staging the data. Once all the PEs have their respective stationary weight elements stored, the input elements are fed from the left edge, with once cycle skew between subsequent rows, as Figure 5.2a shows. At each cycle, each PE would multiply the stored weight element with the incoming input data from the left edge and add it with any incoming partial sum coming from the top edge. The generated result is passed to the neighboring PE via the bottom edge in the next cycle. The skewed delivery of inputs allows the input operand to the bottom PE to arrive at the same time as the accumulated partial sum. In steady state, a reduced full (or partial) output is obtained from at the bottom edge of the array at every cycle, *without any stalls*. Such flow control makes this a systolic array, as discussed earlier in subsection 4.3.3. During the streaming phase, the distribution and collection can occur in parallel requiring only one output buffer at each PE. The accelerator thus supports a distribution and collection bandwidth of 128 GBps. It can also be observed that the PE at the bottom right of the array is the one that produces the last output and determines the overall runtime of the array [78].

Final Design. With the components described above, we arrive at a TPU-like systolic array design shown in subsection 5.3.2a. The 16384 MAC PEs are arranged as a 128×128 square array. The stationary weight elements are fed from the top edge of the array, one element per column every cycle. The streaming elements are fed from the left edge of the array, one element per row every cycle. The final outputs are collected from the bottom edge of the array, one element per column per cycle. A summary of the architectural specs and microarchitecture of buffers and NoC is shown in Figure 5.2b.

Performance with Target Layer. We simulate the example accelerator with the target layer (i.e., the CONV5_2 layer described in Table 5.2) using a Systolic Array simulator called SCALE-sim [78]. The accelerator takes about 58.6 K cycles to finish the computation, at a mapping utilization of 100%. Accounting for stalls when the stationary matrix needs to be replaced, we observe a total throughput of ~1003 MACs/cycle, as shown in as shown in Figure 5.2c. Even though the mapping utilization is 100%, the achieved throughput is about 16× lower than the achievable peak throughput. This is due to the fact that for the target layer dimensions, the data pre-fill and output drain time consume most of the cycles. Specifically, it takes 128 cycles to load the stationary matrix into the array completely, but it remains stationary only for 128 (skew) + 25 (i.e., M) cycles for the target layer, before it needs to be replaced (which we call a *pass*). The number of passes required to compute the full computation is $K/128 \times N/128$ (the number of input tiles across the rows and weight tiles across the columns, respectively) which equals $4608/128 \times 512/128 = 144$.

The key reason for this inefficiency comes from the store-and-forward distribution and reduce-and-forward collection, which scales linearly in such a large design, and becomes the bottleneck at the beginning and end of each pass. To address this, batching can be leveraged in the accelerator, to increase the amount of time the weight matrix stays stationary. With batching, the majority of cycles are spent in performing MAC operations in steady state, and the time spent to fill and drain the array are negligible, which provides a massive boost in achieved throughput to ~14 K MACs/cycle as depicted in the second bar in Figure 5.2c where a batch size of 100 is used for the target layer.

Performance with Alternate Layer. To demonstrate the interdependence of the efficient hardware architecture design and workload parameters, we estimate the performance on the CONV2_2 layer of Resnet50, as depicted in Table 5.2. The runtime for this layer from simulation using SCALE-sim [78] is about 16.5 K cycles. It is worth noting that only 45% of the MAC units were active due to mapping inefficiencies, leaving more than half the array idle. This is due to the fact that the number of filters in the alternate layer is 64 which use only half the columns of the array (unlike 512 filters in the target layer). However, an interesting observation depicted by comparing the first and third bars in Figure 5.2c is that the achieved throughput (without batching) is about 6× more for the alternate layer than the target layer, despite lower mapped utilization. This is due to the fact that the temporal *M* dimension (i.e., the size of each output channel), during which time the weights stay stationary, is higher in the alternate layer than the target layer. This reduces the percentage of time the array stalls during the fills and drains between passes relative to total time. With batching, the throughput goes up to ∼8 K MACs/cycle, but is still significantly lower than that of the target layer—this is where the effect of the mapped under-utilization comes in. All these observations motivate the need for comprehensive design space exploration to make the hardware efficient for as many workloads as possible. We will discuss more on this in Chapter 6.

5.2.2   EDGE DEVICE CLASS EXAMPLE

As the second example, we target edge devices whose area and energy constraints are more stringent than those of data center accelerators. We select MAERI [58] as the target architecture for the example. Figure 5.3 summarizes the architecture whose design flow we discuss next.

Step 1: *Performance Specs.* Assuming a face recognition workload, we target a processing rate of 5 frames per second (FPS) of image classification using Resnet50, which requires five inferences per second. Based on Eyeriss [20], we set the clock frequency to be 200 MHz. Based on the number of operations in Resnet50 (3.6 Billions of MACs), FPS, and the clock frequency, we can compute that the number of MAC operations to be performed per cycle is 89.9. Therefore, we target the number of PEs to be larger than 90. We finalize the number of PEs when we determine mapping to optimize the mapping utilization for the target layer since adding more PEs not utilized by the mapping does not contribute to the performance but only increases the hardware and energy costs.

Step 2: *Dataflow and Mapping.* We select a direct convolution (CONV2D) implementation within the accelerator.

  Parallelized Dimensions. We first parallelize filter height (R) and width (S) dimensions following the strategy of the target architecture, MAERI [58]. This approach results in *virtual neurons (VN)* [58] of size $Sz_{vn} = R \times S = 9$; VNs are groups of adjacent PEs that contribute to distinct output values. The sizes of VNs (i.e., number of PEs per VN) are configurable in MAERI, but in our example we fix the VN size for all layers. Within each VN, each PE (multiplier switch in MAERI) receives different combination of R and S indices since they are parallelized.

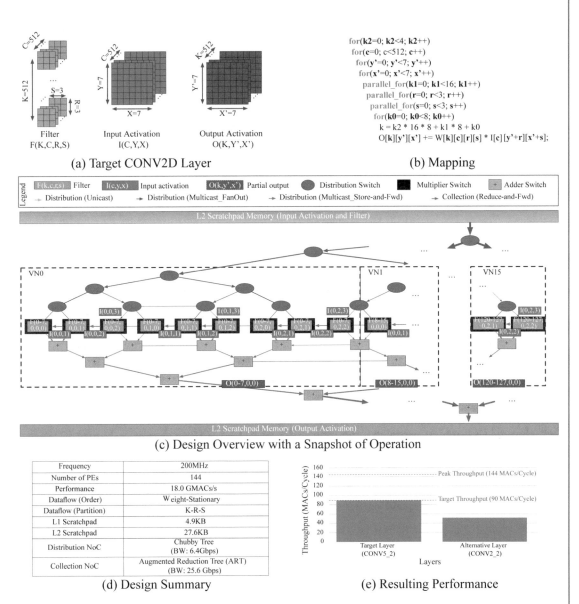

Figure 5.3: An edge device class accelerator based on MAERI [58] designed for CONV5_2 of Resnet50 [5] illustrated in (a). (b) shows the mapping, (c) shows an overview of the design showing a snapshot of a steady state processing, (d) summarizes the final design, and (e) shows the resulting performance of final the design on target and alternative layers we listed in Table 5.2.

To further exploit parallelism (and increase utilization), we select an additional parallelization dimension. Since the target layer in Table 5.2 has a large number of channels compared to
the resolution of activation, we select the output channel dimension (K) for parallelization across
the VNs. Thus, each VN processes different sets of output channels.

Tile Sizing. Considering the parallelization of the output channel (K) dimension and the
VN size, we choose a K tile size that maximizes mapping utilization of the final design. We
assign eight output channels for each VN to reduce the number of filter fetches from L2 to L1
scratchpad. Then, the accelerator covers $N_{vn} \times 8$ (N_{vn}: the number of virtual neurons) output
channels for each computation tile on the entire PE array. We set $N_{vn} \times 8$ to be a positive
integer number that divides the number of output channels to prevent compute unit mapping
underutilization in virtual neuron granularity. Also, we target the total number of PEs, $N_{vn} \times
Sz_{vn} = N_{vn} \times 9$, to be larger than 90 to target the performance goal. To meet both conditions,
we select $N_{vn} = 16$. Based on the choice of N_{vn}, we finalize the number of PEs to be $N_{vn} \times
Sz_{vn} = 144$.

For dimensions other than parallelized dimensions, we select the minimum L2 tile sizes
(tile size of 1 for Y', X', and C dimension, which translates to the tile size of 3 for Y and X
dimensions) to minimize the interconnection bandwidth and buffer size requirements.

Tile Scheduling (Loop Order). We follow the mapping style introduced in MAERI [58].
The filter height (R) and width (S) dimensions are fully unrolled and mapped in each VN, so
those dimensions do not require ordering. We first iterate the output column (X') to exploit the
convolutional reuse supported by the target architecture. After the output column (X'), we iterate over output row (Y') to keep filter values stationary. Then, we iterate input channels (C) to
minimize the buffer size requirements for output activation since the input channel dimension
is accumulated to compute output activation values. Finally, we iterate the last dimension, output channel (K), which completes the loop order of K-C-Y'-X', as shown in Figure 5.3b. The
resulting dataflow is a weight-stationary dataflow because all the dimensions coupled with filter
weight (K and C dimensions) are iterated in the two outer-most loops, which means that the
weight elements are replaced and updated the slowest as compared to the elements of the input
matrix, generated outputs, and partial sums.

Step 3: *Buffer Allocation.* In this step we will determine the sizes of the L1 buffers within the
PEs and outside the array (L2).

L1 Scratchpad Buffer. Based on the mapping described in Figure 5.3b, we can observe
that the tile size of each dimension on L1 buffers in each PE is as follows: (K: 1, C: 1, Y:1, X:1
R:1, S:1). Based on the tile size information and coupled dimensions for each tensor discussed
earlier in Figure 2.11e, we can compute the buffer size requirement as follows:

(1) Filter: $TileSzL1(K) \times TileSzL1(R) \times TileSzL1(S) = 8 \times 1 \times 1 = 8$

(2) Input: $TileSzL1(C) \times TileSzL1(Y) \times TileSzL1(X) = 1 \times 1 \times 1 = 1$

(3) Output: $TileSzL1(K) \times TileSzL1(Y') \times TileSzL1(X') = 8 \times 1 \times 1 = 8$

Considering the double-buffering for latency hiding, we assign 16 buffer slots for filter and output, and two buffer slots for input, resulting in 34-byte L1 buffers. Therefore, the total size of L1 buffers across all the PEs is $144 \times 34 = 4.9$ KB.

L2 Scratchpad Buffer. We apply similar analysis on the L2 buffers as we did for L1 buffers. However, we assume that we allocate enough buffers for the entire non-stationary tensor (input activation in this example), to minimize the number of DRAM accesses. Based on the mapping, we can observe that the tile size of each dimensions on the L2 buffer is as follows: (K:$N_{vn} \times L1TileSz(K) = 16 \times 8 = 128$, C: 1, Y: 3, X: 3, R:3, S:3).

(1) Filter: $TileSzL2(K) \times TileSzL2(R) \times TileSzL2(S) = 128 \times 3 \times 3 = 1152$

(2) Input: $DimSize(C) \times DimSize(Y) \times DimSize(X) = 512 \times 7 \times 7 = 25{,}088$

(3) Output: $TileSzL2(K) \times TileSzL2(Y') \times TileSzL2(X') = 128 \times 1 \times 1 = 128$

Also considering double buffering, the total buffer size requirements is as follows: $1152 \times 2 + 25{,}088 + 128 \times 2 = 27{,}648$, or 27.6 KB.

Step 4: *Network-on-Chip.* During steady states (i.e., when we stream inputs on PEs that already have stationary filter values), we need to deliver three unique input activation values to the PEs from the L2 and collect 16 partial outputs (one from each VN), every cycle. We use a chubby tree (described earlier in Section 4.3.1) for distribution from the L2 to the multipliers, and set the root (i.e., L2 read) bandwidth to be four words (to keep it a power of two). The same three input activations are spatially reused by all VNs via multicast. The other six activation values within the convolution window are distributed via store-and-forward within each VN for exploiting convolutional reuse (input activation overlap across adjacent sliding windows). For collection, we use the ART topology, described earlier in Section 4.3.1, and set its root (i.e., L2 write) bandwidth to be 16 words. This translates to 6.4 Gbps and 25.6 Gbps NoC bandwidth based on our clock frequency and data precision assumptions. Given this bandwidth, the PEs can operate in a stall-free manner during steady state. However, they would need to stall when the stationary tiles are being replaced. Each of the PEs (multipliers/adders) employs a latency-insensitive flow-control mechanism, as discussed earlier in subsection 4.3.3. This means that the accelerator can operate correctly (albeit at lower throughput) even if the required distribution and/or collection bandwidth is more than the one provisioned in the NoC (which may be the case for other layers mapped on the accelerator).

Final Design. The final 144-PE MAERI design is shown in Figure 5.3c. It operates as follows. First the accelerator is configured with a VN size of 9 (as discussed above). Next, the stationary filter tiles are distributed to all the multipliers from the L2. The input activations are then streamed through the distribution NoC. In steady state, three new activations are streamed to each VN, and the rest of the six are forwarded locally within each VN. In each cycle, all the multipliers in each VN produce partial sums, that are then spatially accumulated via the ART.

Each of the 16 VNs produce one full (or partial) output that is sent to the L2. A summary of the architectural specs and microarchitecture of buffers and NoC is shown in Figure 5.3d.

Performance with Target Layer. We analyzed the final design using MAESTRO [79]. Processing the target layer (i.e., the CONV5_2 layer described in Table 5.2) requires 661.6 K cycles, which translates to a latency of 3.1 ms. During steady state, the mapping utilization on the target layer is 100% leading to 144 MACs/cycle. However, when a new set of weights are fetched in the next tile, 1152 distinct weight values need to be distributed from the global buffer to PE array, which leads to 288 cycles of stalls since the communication is not completely hidden behind the compute. This leads to an overall effective throughput of 89.2 MACs/cycle in the accelerator, as shown in Figure 5.3e, which is sufficient to meet the target performance specs of 5 fps at 200 MHz.

Performance with Alternate Layer. A hardware design tuned for one layer may not always perform well on other layers. This is evident when we run a different layer, CONV2_2 layer of Resnet50 [5], whose dimensions are shown in Table 5.2, on the same architecture and evaluate the performance using MAESTRO. The new layer has relatively smaller number of channels and higher resolution of input activations compared to the optimization target layer in Figure 5.3a. Note that although MAERI can adapt to new layers by changing the mapping and configuring the VN sizes accordingly, we use the same mapping and VN configuration as the ones used for the target layer to isolate the impact of using the same microarchitecture. The resulting number of cycles is 2.1M, translated into the latency of 10.4 ms. The mapping utilization on the new layer decreases to 50% since 8 VNs will cover the entire output channel dimension ($TileSzL1(K) \times 8 = 64$ channels) and the other 8 VNs are not utilized. This leads to a severe throughput drop, as shown in Figure 5.3e. The throughput is 51.8 MACs/cycle, which is 42.0% lower than that of running the original target layer. Such results suggest that each layer prefers different hardware configurations, and design-space and map-space explorations are crucial, as we discuss in Chapter 6.

5.3   CASE STUDIES

In Table 5.3, we compare six DNN accelerator prototypes across the topics discussed in the book so far (dataflow, buffer hierarchies, and on-chip networks). We picked Eyeriss [64], TPU [25], NVDLA [26], MAERI [58], Eyeriss v2 [69], and SIMBA [61], since they offer contrasting implementations. The purpose of this exercise is not to endorse any specific dataflow, architectural or implementation decisions, but to enumerate the choices.

5.3.1   EYERISS (2016)

Eyeriss [64] is a 168-PE accelerator developed and prototyped at MIT.

Dataflow. Eyeriss employs a Row-Stationary dataflow strategy. The rows from each filter are mapped spatially along the PEs in each column. They remain stationary. They are replicated

Table 5.3: Comparison of data orchestration mechanisms within DNN accelerators

		Eyeriss (2016)	TPU (2017)	NVDLA (2017)	MAERI (2018)	Eyeriss v2 (2019)	SIMBA (2019)
	Computation	CONV	GEMM	CONV	CONV	CONV	CONV
Dataflow	**Innermost Loop Order**	RS	WS	WS	Flex	RS	WS
	Partitioning	Y-R	K-C	K-C	Flex	Flex	K-C
Buffer Hierarchy	**Orchestration Approach**	EDDO	EDDO	EDDO	EDDO	EDDO	EDDO
	Buffer Implementation	Custom + FIFO	Custom	Custom	FIFO	Custom + FIFO	Buffet
	Buffer Partitioning	L1-Hard L2-Soft	L1-Hard (Latch) L2-Soft	L1-Hard (Latch) L2-Hard (Out) L2-Soft (I + W)	L1-Hard L2-Soft	L1-Hard L2-Soft	L1-Hard L2-Soft
On-Chip Network	**Distribution (Input Act)**	Multicast via Bus	Multicast via Store-Fwd	Unicast via Bus	Multicast via Tree	Multicast via H-Mesh	Multicast via NoC+ NoP
	Distribution (Weight)	Multicast via Bus	Unicast via Fwd	Multicast via Bus	Multicast via Tree + Store-Fwd at Leaves	Multicast via H-Mesh	Multicast via NoC+ NoP
	Collection (Output)	Reduce-Fwd, then Bus to L2 Buffer	Reduce-Fwd, then RMW Buffer	Reduce via Tree	Reduce via Augm Redn Tree	Reduce-Fwd, then HMesh to L2	Reduce via Tree till L2, then unicast + L2 RMW
	Link Partitioning	All Dedic	I+W: Shared Out: Dedic	I+W: Shared Out: Dedic	I+W: Shared Out: Dedic	I+W: Shared Out: Dedic	All Shared over NoC and NoP

across the columns. Rows from the input are streamed into the array and move in a diagonal manner, to implement convolutional reuse. Each column produces an output activation.

Buffer Hierarchy. Each PE houses a local private scratchpad to hold the stationary weight rows, input rows, and output partial sums. There is a global on-chip SRAM, shared by the entire array. Within the global buffer, the output banks are hard partitioned, while the rest of the banks can be shared by weights and inputs via soft partitions.

Network-on-Chip. For distribution of weights and inputs, Eyeriss uses separate hierarchical buses. Weights are multicast horizontally per row, while diagonals are multicast diagonally. Each PE has an id which is used to determine whether the operands on the broadcast buses should be consumed by a PE or not. Eyeriss allows PE ids to be configurable when mapping the algorithm—thus allowing it to mimic arrays of different aspect ratios. Reduction of partial sums happens via reduce-and-forward links along each column. The final outputs from each column use a collection NoC, implemented as a hierarchical bus like the distribution NoC, to deliver outputs back to the global buffer.

5.3.2   TPU (2017)

The TPU [25] is a commercial accelerator deployed in Google's datacenters. TPU v1 housed 256×256 8b fixed point PEs, while v2 and v3 changed it to 128×128 arrays with floating point PEs [80].

Dataflow. The TPU is a weight-stationary systolic array. It runs GEMMs instead of CONV. Each weight filter is converted into a vector and fed along each column. It remains stationary. Each row feeds the flattened input. Each PE produces a partial sum every cycle, and the partial sums along a column are summed up for the final output.

Buffer Hierarchy. Each "PE" in the systolic array is a simple MAC unit with one latch for the stationary weight operand, and another latch for the input operand received from the neighbor. There is no other buffering within each PE. The perimeter of the array houses a multi-banked L2 SRAM. Each row of the TPU needs access to its own SRAM bank, to maintain full throughput. In addition, each column has an accumulation buffer for accumulating partial sums into an output. It is used in cases when the size of the filter is larger than 128 (the length of the column).

Network-on-Chip. The TPU uses store-and-forward for unicast distribution of weights, and multicast distribution of inputs. It uses reduce-and-forward for reduction and collection across the columns. There are additional accumulation buffers at the ends of each column to perform a RMW temporal reduction if the entire dot-product does not map over the column in one pass.

5.3.3   NVDLA (2017)

NVDLA [26] is an open-source accelerator developed by NVIDIA, and part of its Xavier SoC.

Dataflow. NVDLA employs a weight-stationary dataflow. PEs are grouped into *Atomic-K* clusters of size *Atomic-C*. These names represent the dimensions being parallelized. Within each cluster, the parallelization is across the input channels, while across clusters the parallelization is across the output channels.

Buffer Hierarchy. Each PE holds one weight stationary. All clusters share a global L2 buffer. Within the L2 buffer, the output banks are hard partitioned, while the rest of the banks can be shared by weights and inputs via soft partitions.

Network-on-Chip. Input and Weight Data distribution in NVDLA occurs over buses. The partial sums within each cluster are reduced spatially using a tree. The outputs from different clusters directly connect to the output SRAM banks.

5.3.4   MAERI (2018)

MAERI [58] is an open-source reconfigurable accelerator substrate developed at Georgia Tech. It targets extreme flexibility in supporting arbitrary dataflows. It uses a logically linear chain of multipliers instead of a rigid 2D array.

Dataflow. MAERI allows any dimension/tensor to remain stationary. The parallelization strategy is also left to the mapper. The granularity of parallelization is based on a configurable parameter called virtual neuron (VN). All PEs within a VN compute the same output. But the size of the VN is completely configurable. For instance, a possible dataflow is to keep weights stationary at the multipliers, unroll the channels, and this determines the VN size. Across VNs, different output channels may be mapped.

Buffer Hierarchy. MAERI uses FIFOs at each multiplier for the operands. It relies on in-order delivery of operands through the distribution network, and therefore does not need addressable scratchpads. There is a shared global SRAM scratchpad called prefetch buffer for the entire array.

Network-on-Chip. MAERI uses a fat binary-tree based topology for distributing inputs and weights. Depending on the dataflow strategy, the inputs and/or weights could be multicast. The bandwidth at the root of the tree depends on the read bandwidth from the prefetch buffer. To implement convolutional reuse, the inputs distributed through the tree are then forwarded cycle by cycle between the multipliers within a VN. Collection of outputs occurs via a separate fat binary-tree. This tree is called Augmented Reduction Tree (ART) and uses additional forwarding links to support non-blocking reductions. Each VN performs a tree-based reduction, and the final outputs bypass the adders on the tree to get written back to the prefetch buffer. The bandwidth at the root of the ART depends on the amount of write bandwidth at the prefetch buffer.

5.3.5   EYERISS V2 (2019)

Eyeriss v2 [69] is a follow-on work to Eyeriss. This accelerator also supports weight sparsity.

Dataflow. Eyeriss v2 enhances the Row-Stationary (RS) dataflow used in the original Eyeriss. It supports spatial mapping of any layer dimension, and leverages this for mapping channel group dimensions that occur in depth-wise convolutions spatially.

Buffer Hierarchy. Eyeriss v2 has a similar 2-level hierarchy like Eyeriss. However, the global buffer in Eyeriss v2 is distributed across PE clusters, instead of being one monolithic block at one end.

Network-on-Chip. Eyeriss v2 introduces a novel Hierarchical Mesh topology, optimized for multicast-based distribution of inputs and weights to PE clusters. It operates in multiple modes: (i) high bandwidth mode; (ii) high reuse mode; (iii) grouped-multicast mode; and (iv) interleaved-multicast mode, depending on the partitioning strategy employed within the dataflow. Within each PE cluster, the PEs use a bus for distribution. Reduction within a cluster occurs via reduce-and-forward.

5.3.6   SIMBA (2019)

SIMBA [61] is a scale-out Multi-chip Module (MCM) DNN accelerator prototyped by NVIDIA. It is built using multiple individual DNN chiplets connected via a network-on-package (NoP).

Dataflow. SIMBA uses weight-stationary dataflow at the individual MACs. For parallelization, it leverages input channel partitioning within a chiplet, and output channel partitioning across the chiplets. It also supports hybrid partitioning strategies.

Buffer Hierarchy. SIMBA employs a two level memory hierarchy within each chiplet. Across the chiplets, the cache residing on the host CPU is shared for exchange of information.

Network-on-Chip. SIMBA uses a conventional mesh-based NoC within each chiplet, and a mesh-based NoP across chiplets. Each chiplet houses the NoC and one NoP router. All data distribution and collection happens in a packet-switched manner over the NoC and NoP.

5.4   SUMMARY

In this chapter, we discussed a systematic process for DNN accelerator design and demonstrated this flow with two cases studies. We also presented and contrasted architectural design choices across a suite of recent DNN accelerators. This illustrates how large and complex the design space is. How can the architect be sure they are not missing out on a significantly better design simply by moving slightly in the parameter space? In the next chapter, we present mechanisms to quantify the cost of various design choices, and discuss tools to aid in the automation of this complex process.

CHAPTER 6

Modeling Accelerator Design Space

Modeling performance and energy efficiency of designs via cycle-level simulation has been an indispensable tool for architectural design space exploration, especially in the traditional CPU space. The simplicity and regularity of DNN accelerator architectures allows for commensurately simple (and therefore fast) high-fidelity models. However, unlike traditional CPUs, DNN accelerators do not have a standard ISA with pre-compiled workloads that work on a variety of designs with the same ISA. Instead, the storage and networking hierarchies of the design are exposed to the programmer, and each workload must be optimally mapped (compiled) to take full advantage of the attributes of any new design under consideration. This makes mapping an essential step in the design-space exploration loop for DNN accelerators. In this chapter, we discuss the mapping step in further detail, and then examine how microarchitectural models for DNN accelerators can be constructed to evaluate the performance and energy cost of execution of a mapping on the hardware.

6.1 SEPARATING THE MAPPING SPACE FROM THE ARCHITECTURE DESIGN SPACE

As we saw in Chapter 2, the vast number of possible dataflows and the accompanying choices of hardware implementation results in a large *architecture design space*. In contrast, for a given DNN layer, i.e., a *workload*, a flexible architecture may permit many different ways of scheduling operations and staging data on the same architecture, which we call different *mappings* [81], resulting in widely varying performance and energy efficiency. Therefore, to characterize the performance and energy efficiency of a workload on an architecture, we need to search for a good mapping within the space of valid mappings, i.e., the *mapspace*. The need to do this search for each workload compounds the complexity of exploring the architecture design space.

DNN accelerator proposals address this mapping problem within the scope of their specific designs [20, 58, 74, 82–87]. In contrast, tools for broader design space exploration [46, 83, 88–90] must grapple with the more challenging problems of separation of the architecture design space from the mapspace, automatic generation of the mapspace given a workload and an architecture, and heuristics to search through the mapspace.

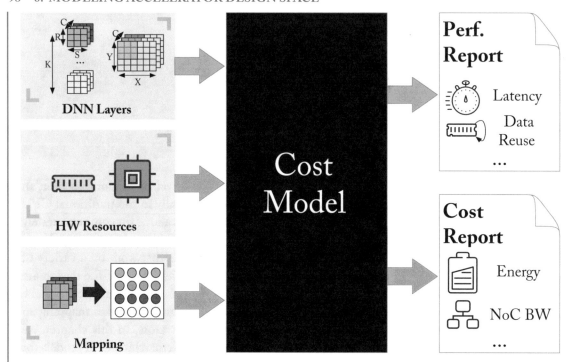

Figure 6.1: The role of a cost model.

As a specific example, Timeloop [88] approaches the problem by separating the infrastructure into two main components: a *model*[1] to provide performance, area and energy projections and a *mapper* to construct and search through the mapspace of any given workload on the targeted architecture. There exists a unique symbiosis between a model that can describe and emulate a wide variety of architectures and a mapper that can optimally configure the accelerator to get a fair characterization of the performance and energy efficiency of a specific workload.

A model needs a mapper. Unlike traditional architectures that use an ISA as the hardware/software interface, each DNN accelerator uniquely exposes many configurable settings in the hardware, and its behavior is critically dependent on those settings. Therefore, instead of using pre-compiled binaries or traces, running a DNN workload on an accelerator involves finding a mapping (and corresponding set of configuration settings), which dictates the scheduling of operations and data movement, to maximize performance and energy efficiency.

A mapper needs a model. A mapper enumerates the mapspace and then iteratively searches through the space for the optimal mapping. Thus, it needs a cost function to determine whether one mapping is better than another. This cost function must be *fast* enough to

---

[1]The word model in this chapter refers to the microarchitectural model of the accelerator, not the DNN model, which is referred to as workload.

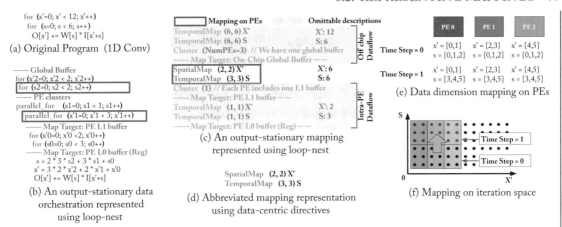

Figure 6.2: A mapping description example over a 1D convolution program described in (a). (b) and (c) show loop nest (compute-centric) and MAESTRO IR (data-centric) description of mapping depicted in (e) and (f). (d) shows an abbreviated version of data-centric description given in (c).

evaluate to allow the mapper to explore a large mapspace in a reasonable amount of time, but it must also be *accurate*, i.e., it must be representative of the actual behavior of the mapping on the architecture, otherwise the mapper will give misleading results.

6.2   REPRESENTING MAPPINGS

We refer to the architecture-independent algorithmic specification of a problem (as in Figure 6.2a) as an *unmapped* representation. The set of dynamic instances of the loop body in this unmapped representation forms the *iteration space* of the workload. Each dynamic instance of the body includes one or more references to a set of operand/result tensors using affine indexing expressions over loop variables. We call these tensors *dataspaces*. The set of dataspace points that are accessed by a set of iteration space points can be computed by *projecting* the coordinates of the iteration space points into dataspace coordinates using the indexing expressions.

An architecture-specific *mapping* describes the way in which the iteration space and the associated dataspaces are split into *tiles* (chunks) at each level of the memory hierarchy and among multiple memory instances at each level.

More precisely, a mapping identifies the point(s) within the iteration and data spaces that are being worked on or are resident at hardware instance i and time t. While the complete set of legal mappings of a workload onto an architecture is at least countably infinite, prior works have used stylistic ways to concisely describe a *useful* subset of mappings. These stylistic approaches not only allow mappings to be described in an intuitive, human-readable manner, but also allow for programmatic exploration of the space of mappings. We now discuss two of these approaches.

6.2.1   LOOP NEST BASED MAPPING SPECIFICATION

A *loop nest* based mapping representation can describe all facets of a mapping in a single unified format [88].

Figure 6.2a shows a 1D convolution workload (introduced in Section 2.2) with input activation of size 12 and filter size 6, represented as an *unmapped* loop nest [88]. Figure 6.2b shows an example *mapped* loop-nest targeting an architecture consisting of an array of 3 PEs, each with a private MAC unit, an L0 register and a private L1 buffer. All PEs share a global buffer.

To construct the mapping, the 1D convolution loop nest is split into a number of sections (called *tiling levels*) equal to the number of storage hierarchy levels, plus the number of levels with spatial fanouts. Each tiling level has a loop corresponding to each dimension in the original workload (though the bound may be 1). The product of all the loop bounds belonging to a dimension must be equal to the final (optionally padded) value of the dimension.

Using the mapping in Figure 6.2b as an example, we observe the following.

(1) Loop bounds at each tiling level determine the size of the tile for each dataspace held at that level. For example, each PE's L1 buffer owns an $3 \times 2 = 6$-sized iteration space tile, which projects onto 3, $2 + 3 - 1 = 5$, and 2-sized weight, input, and output tiles, respectively. Size of the dataspace tiles is constrained by the size of the buffer at each level.

(2) `parallel_for` loops represent spatial partitioning of tiles across instances of a level. For example, the `parallel_for` loops distribute the 1×3 iteration space (and the dataspace projections thereof) across 1×3 PE instances. This specific mapping results in replication of some input data (called a *halo*—Section 2.2) between adjacent PEs.

(3) Ordering of loops within a tiling level determines the sequence in which sub-tiles will be delivered from that level to an inner level during execution. For example, iterating over the $s2$ loop at the global buffer results in interesting changes in each PE's dataspace tile: the outputs remain *stationary*, the weights get replaced each iteration, and the inputs display an overlapping *sliding-window* pattern, which means the global buffer only needs to supply the portion of the tile that is not already present in the PE. Inferring these dataspace changes (or *deltas*) is a primary function of the tile-analysis process described in subsection 6.3.1.

This representation produces a strictly inclusive tile hierarchy, which may not be optimal. At times, allowing a dataspace to *bypass* a level opens up the capacity to other dataspaces, enabling larger tile sizes and potentially resulting in a more optimal mapping. Representing bypassing requires an auxiliary directive that can be specified for each tiling level in the loop nest. This directive specifies which dataspaces are allowed to reside at each level.

The approach used for this 1D convolution example can be easily extended to a higher-dimensional space, such as a full CONV2D layer. Such a mapping representation allows us to reason about the space of possible mappings (or *mapspace*) for an architecture in a structured, programmatic manner.

Directive Syntax	Semantics (When used over dimension 'A')	Example
Temporal_Map (Sz, Ofs)	Cluster $_{targ}$[idx](A) <- {**a** | ai * Ofs <= **a** < ai * Ofs + Sz} - ai is the iterator for dimension A, which represents the number of temporal iteration of A.	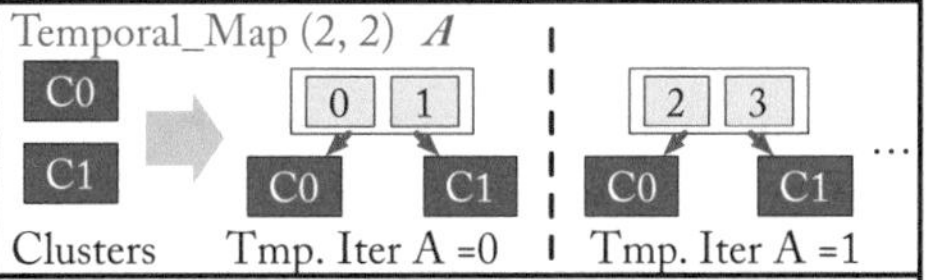
Spatial_Map (Sz, Ofs)	BaseIdx = (ai * NumClusters) * Ofs + ClusterIdx * Ofs Cluster[ClusterIdx](A) <- {**a** | BaseIdx <= a < BaseIdx + Sz} - NumClusters is the number of clusters at this spatial map directive is specified. If no cluster directive is used, it is the same as the number of PEs.	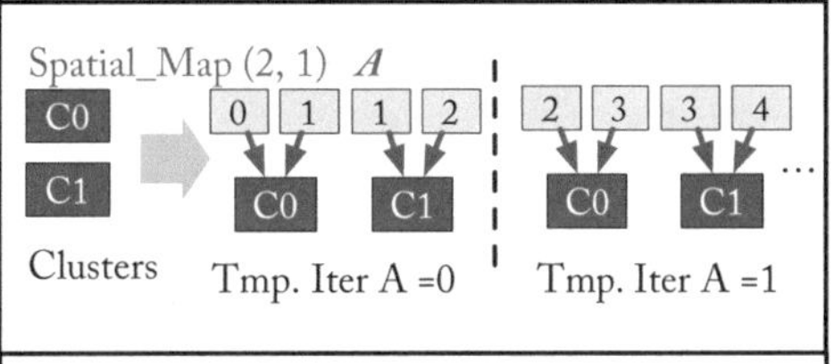
Cluster (Sz, Cls)	Cluster $_{new}$[Idx$_{new}$] = {Cluster$_{old}$[Idx$_{old}$] | Idx$_{new}$ = Idx $_{old}$ / Sz} - Clusters are recursively constructed from the inner-most cluster directive to the top-most one. - Mapping directives above and below cluster directive applies on Cluster $_{new}$ and Cluster $_{old}$, respectively. - Cluster class (Cls) can be either *Logical* or *Physical*. If a cluster is physical, all the sub clusters are mapped on a physical PE or a PE cluster - If no physical cluster is specified, the inner-most cluster is mapped on a physical PE	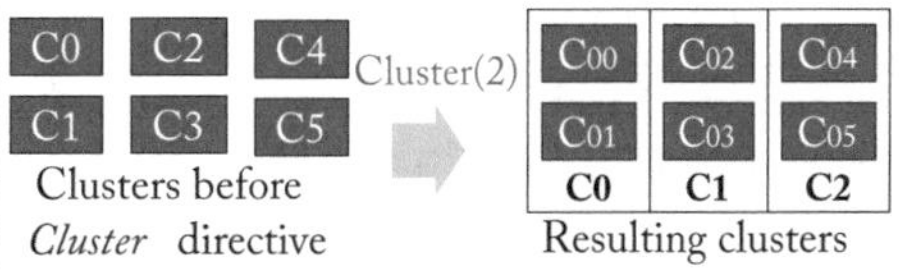

Figure 6.3: Three MAESTRO data orchestration directives and their semantics.

6.2.2 DATA-CENTRIC MAPPING SPECIFICATION

While a loop nest based (or *compute-centric*) mapping representation is sufficiently precise to deterministically derive tile placement and data movement, a *data-centric* mapping description directly describes data tile mapping and movement. An example is the MAESTRO [79] data orchestration IR. As shown in Figure 6.3, the MAESTRO data orchestration IR has three primitives: *TemporalMap*, *SpatialMap*, and *Cluster*. Each of the map directives specifies data tile movement over temporal dimension (time) and spatial dimension (PEs or PE clusters). The cluster directive allows recursive construction of PE clusters (e.g., 72 PEs -> (applying cluster (3)) 24 clusters of 3 PEs -> (applying cluster (2) -> 2 clusters of 12 clusters of 3 PEs.). By doing so, MAESTRO allows arbitrary granularity of mapping over an accelerator with arbitrary PE array dimensionality.

Figure 6.2c shows the data-centric representation of the same mapping that is expressed as a loop nest in (b). The major difference between two representations is that compute-centric version in (b) describes the number of tile iterations while the data-centric version describes the tile size and how the tile moves via offset, as shown in red boxes in (b) and (c). Also, MAESTRO's data-centric description can be written in a concise format as shown in (d) while preserving all the crucial mapping information. For example, the inner-PE dataflow parts, tile sizes of each dimension are typically one unless the PE has a SIMD style wide-vector ALU, which is not typically treated as one PE. Also, the impact of ordering is minimal within a PE since all the data points in a data tile on a PE are kept local in a PE. This enables a cost model (described in the next section) to automatically infer one of the possible data orchestration within a data tile on a PE so that users can focus on impactful part of the mapping.

6.3   MODELING THE EXECUTION OF A MAPPING ON AN ARCHITECTURE

Given an architecture specification, a problem specification and a particular mapping, the job of a model is to analyze the behavior of the mapping on the hardware and estimate cost metrics such as performance and energy efficiency.

Recall that a mapping is a precise specification of the partitioning of iterations and data across all hardware units and sequencing of data and operations at each unit over time. Thus, classic modeling techniques such as microarchitectural cycle-level modeling or RTL simulation can be used. However, these approaches are too slow to be useful as in-line cost models for a mapper.

Fortunately, the regularity of DNN workloads allows for dramatically faster *analytical* approaches. These approaches provide similar accuracy to detailed cycle-level models but are orders of magnitude faster than conventional simulators (typically faster than running the workload natively on contemporary hardware). The analysis process in these approaches is typically structured into the following phases.

1. A **tile analysis** phase determines the hierarchical *tiles* of data that represent the mapping and measures the transfer of data between them to obtain *tile-access* counts.

2. A **microarchitecture model** uses these tile-access counts to derive the access counts to hardware components, which can be used to derive performance.

3. A **technology model** provides technology-specific area and energy data, which in combination with the hardware specifications and modeled microarchitectural statistics can provide area and energy estimates.

6.3.1   TILE ANALYSIS

Analyzing a Mapped Loop Nest

As explained in Section 6.2, loop bounds at each tiling level determine the tile for the iteration space, and in turn for each dataspace held at that level. To compute *tile-access* counts, we must determine the volumes of data that must move between the tiles over space and time to execute the schedule dictated by the mapping. To perform this analysis, the model needs to track the set of points representing tiles of the iteration and operand/result spaces at each time step and at each instance of a hardware unit (such as a multiplier or a buffer instance).

Consider two consecutive iterations i and $i + 1$ at any loop in a mapping (from Figure 6.2). At each iteration, we can determine the point sets required by the complete sub-nest underneath this loop. We call the set-difference between these point sets a *delta*. The size of this delta has different connotations for spatial and temporal loops.

For a temporal (`for`) loop, an empty delta indicates perfect reuse over time (*stationarity*), requiring no additional data movement. A non-empty delta—indicating no reuse or partial reuse (such as in a sliding-window pattern)—represents the incremental data that must be transferred between levels.

For a spatial (`parallel_for`) loop, the deltas represent overlaps between data held at adjacent hardware instances (represented by the iterations i and $i + 1$). An empty delta at the same time step indicates a *multicasting* opportunity, while an empty delta at consecutive time steps between adjacent hardware instances indicates a *forwarding* opportunity (such as in a systolic array).

To compute all tile access counts, we need to measure and accumulate deltas over all space (all hardware units) and time (the complete execution of the mapping). A naïve but robust way to do this is to effectively *simulate* the execution of the entire loop nest on the architecture. Sadly, this is unacceptably slow. Fortunately, DNN workloads index into operand and result tensors in a very regular way, and in most cases with each loop index variable showing up only once in the indexing expressions for each tensor. This allows for two optimizations. First, the model only needs to compute the tiles for the first, second and last iterations of each loop. The deltas between other iterations will be of the same shape and size as between the first and second iterations, so the access costs of these other iterations can be algebraically extrapolated. Second, each tile shape is an *axis-aligned hyper-rectangle* within the tensor, which allows for easy delta calculation.

Analyzing a Data-Centric Mapping Representation

Similar to the loop-nest based delta-analysis described above, analyzing a data-centric description also involves computing deltas to identify the data reuse. However, starting from a data-centric description as an input intermediate representation (IR) simplifies the process and reduces the estimation time for each run.

As shown in Figure 6.4, delta can be simply computed as follows: offset of *changing* dimension multiplied with mapping size of all the other dimensions in a data class, as an example in Figure 6.4 shows. The same method can be applied to both temporal and spatial map (corresponds temporal for and parallel for in loop nest representation). To analze data reuse over entire execution, we need to further identify (1) how the dimensions are coupled with each data

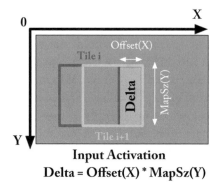

<Data Orchestration>
TemporalMap(MapSz(Y), Offset(Y)) *Y*
TemporalMap(MapSz(X), Offset(X)) *X*

Input Activation
Delta = Offset(X) * MapSz(Y)

Figure 6.4: Delta volume computation for reuse analysis in MAESTRO.

Variable / DataClass	Output Channel (K)	Input Channel (C)	Filter Row (R)	Filter Column (S)	Input Row (Y)	Input Column (X)
Output Activation	×		×	×	×	×
Input Activation		×			×	×
Filter Weights	×	×	×	×		

* Output row (Y') = Y-R+1, Output column (X') = X-S+1

Figure 6.5: Coupled dimensions for each data class (tensor) in conv2D operation. "X" represents coupling.

class (filter and input/output activations), (2) which dimension changes over tile movements, and (3) how many times each tile movement pattern is repeated.

Dimension coupling   The first consideration is dimension coupling with each data class, or tensors. Figure 6.5 shows the data coupling of full convolution, conv2D operator, one of the most common operations in many DNNs. Coupling affects if a data tile of a tensor changes or not under a mapping update along one of the dimensions. For example, when data mapping of output channel (K) is updated from tile j to tile j+1, input activation is *stationary*, which implies a temporal reuse opportunity via buffers, since input activation is not coupled with output channel dimension so input activation tile does not change. When those two tiles are spatially mapped, it implies a *multicasting* opportunity, or a spatial reuse opportunity via NoC, since input activation tile is identical across hardware instances. As the example reveals, the dimension coupling relationship is the essence of data reuse opportunities in DNNs; when a dimension changes, certain data tile of a tensor might stay identical based on the dimension coupling, which implies reuse. To analyze data reuse over entire execution based on this key insight, we need to identify tile movement patterns per dimension and changing dimensions in each pattern.

Tile movement patterns   Although the delta-based analysis discussed above is effective, such an analysis only covers "steady" iteration states in the most fine-grained tile iterations, or the inner-most loop iterations in loop nests. That is, the analysis is not valid in the first and last iteration of each set of inner-most loop iterations. For example, the 1D convolution data orchestration example in Figure 6.2, when the tile covers entire filter (S) values at time step 1 as shown in (f), than the tile will move on to the next set of output (X') values, (6,7), (8,9), and (10,11) on PE0, PE1, and PE2, respectively, at time step 2. At the transition of the tile from time step 1 to 2, the mapped filter values will be initialized, or "reset," to the first mapping. Such an initialization is not based on offset so delta is the entire filter tile in this case. Also, if the filter dimension was 5 (instead of 6), the last iteration will have a smaller number of mapped data elements—two filter values—even though the delta-based analysis indicates three mapped data elements. In full convolution (conv2D operator) with six dimensions (seven when we include input batch), such exceptional cases occur on each dimension. To analyze such cases, the model lists up all the possible iteration states, which is a vector of iteration states over each dimension (e.g., (K=init, C=init, R=init, S=init, Y=steady, X=steady)) and identifies the number of each case through the entire computation. Based on the coupled variable analysis, the model identifies data reuse of each data class and computes a set of data transfer counts. These counts can then be applied to a microarchitecture model (discussed next) to derive activity counts such as the number of buffer accesses, MACs, NoC transfers, and so on.

6.3.2   ANALYTICAL MICROARCHITECTURE MODEL

The simplicity of DNN accelerator microarchitectures (in comparison to, say, a high-performance CPU microarchitecture) allows them to be modeled at a somewhat higher level of abstraction than traditional CPU/GPU microarchitecture models without sacrificing fidelity. The hardware organization is typically represented as a tree [79, 88] with buffers, arithmetic units, and state machines as nodes and networks/wires as edges. Each node, such as a scratch-pad buffer, may be modeled at varying levels of fidelity. For example, SRAM buffers may be modeled abstractly as a size vs. energy-per-access function. This will probably provide imprecise projections for a specific architecture, but may provide architects with intuition on how energy scaling trends interplay with tile reuse benefits and affect overall energy efficiency. At the other extreme, one can model SRAM buffers with an extensive database of pre-generated SRAM macros in specific technologies with various aspect ratios and bank configurations. Generating such a database requires extensive effort and access to proprietary technology information and is also tedious to parameterize correctly by a user, but is capable of generating more precise projections. The choice of fidelity must be well-matched with the level of detail captured by the data transfer activity counts derived during the tile-analysis (subsection 6.3.1) phase.

Estimating Performance

Because of the regular computation and data movement patterns exhibited by DNN layers, buffer accesses and network transfers can usually be statically scheduled, largely avoiding irregular and disruptive bank and network conflicts. Therefore, given a set of tile-access counts, performance can be projected using a throughput/bandwidth based approach, by first calculating the number of cycles it would take for each hardware component to complete the workload in isolation.

For arithmetic units (such as MAC units), this isolated execution cycle count is equal to the number of arithmetic operations in the workload divided by the number of hardware units.

For communication interfaces, the isolated cycle count is equal to the amount of data flowing through that interface divided by its bandwidth. For example, consider an architecture with a Global buffer supplying data to an array of N PEs over a custom multicast network. Tile analysis (subsection 6.3.1) of a specific mapping may tell us that over the course of execution of the entire mapping.

- D bytes of data are read from the Global buffer and distributed (i.e., no multicast) to the N PEs.

- E bytes of data are read from the Global buffer and broadcast to the N PEs.

From this information, we can derive the following.

- The isolated execution cycle count of the read-port of the Global buffer is equal to $(D + E)/R$ where R is the read bandwidth provided by the port.

- The isolated execution cycle count of the fill-port of each PE's local buffer is equal to $(D/N + E)/F$ where F is the fill bandwidth provided by the port.

- The isolated execution cycle count of the network is $(D + E)/B$ where B is the injection bandwidth allowed by the network at the Global buffer end (assuming it has sufficient internal bandwidth to deliver all multicast/distribution patterns to support the injection rate).

This was a simple example using a high-level stateless model, but it illustrates how data-movement statistics from the tile-analysis phase can be translated into performance projections.

Next, we need to integrate these isolated cycle counts into an overall performance projection. In most hardware accelerator architectures, buffers, networks and arithmetic units are staged into a pipeline. Thus, the overall execution latency can be estimated as the maximum of isolated execution cycles across all buffers, networks, and arithmetic units in the hardware, plus an initiation latency to fill the entire pipeline. Such a stateless throughput-based model [79, 88], which assumes negligible pipeline stalls, is reasonable for architectures that use double-buffering or *buffet*-based pre-buffering (Chapter 3).

Hardware Unit Access Counts for Energy Estimation

Similar to performance, energy can be estimated by first transforming data transfer counts obtained from the tile-analysis step (subsection 6.3.1) into access counts for various microarchitectural structures.

Arithmetic unit activity, buffer accesses and network ingresses are linear functions of data transfer counts between levels obtained from tile analysis. For buffers, microarchitectural parameters such as number of banks, aspect ratio, etc., must be factored into the access count calculations. Network activity includes traffic between levels (including fan-out and multi-cast traffic), as well as traffic between storage instances at the same level (if they forward data to each other). Multi-cast signatures from the tile analysis stage are used to determine the spatial locations of child instances targeted by a specific parent→child data transfer. The architectural topology determines the number of hops required for the data to be routed to the destinations. Additionally, a number of architecture-specific attributes can be factored into a linear analytical model. For example:

- some architectures may elide the first zero-read, and may support local accumulation at some buffers;

- some architectures support *spatial reduction* of output tensors; and

- some architectures have dedicated *address generators*, which are adders surrounded by state machines to generate read and write addresses for storage elements.

These hardware unit activity/access counts then need to be combined with a per-access energy cost to determine total energy consumption for the workload. These energy costs are often derived from technology-specific models, which we describe next.

6.3.3   TECHNOLOGY-SPECIFIC ENERGY/AREA MODELS

While performance can be derived directly from microarchitectural statistics, estimating energy consumption requires access to technology-specific data on energy cost to access various hardware structures. Because these energy costs are often data-dependent, precise energy estimation requires at least cycle-accurate simulation with realistic data patterns. However, it has been shown that simple analytical approaches, such as multiplying the hardware component access count (taking sparsity into account) with an *average* energy-per-access is reasonably precise for DNN workloads [88].

Unfortunately, even obtaining these average energy-per-access costs is non-trivial. Costs are required for various hardware structures—RAMs, register files, arithmetic units, on-chip network routers, and data wires. Costs vary with technology nodes, and they do not scale with technology in consistent ways. Furthermore, cost data is often proprietary may only be available for a coarse subset of design points.

These are fundamental challenges that are difficult to overcome, but recent works [91] attempt to streamline the process of using whatever technology data may be available to a user by providing unified but extensible interfaces to data points from a diverse set of sources. Such technology-specific area/energy models [88, 91] are usually built up from a number of key subcomponents.

(1) A *Memory Model*, which supports modeling of memories with different sizes, aspect ratios, number of ports and banks. It also supports different memory implementations, including DRAM, SRAM, and register files. The SRAM and register-file component can be based on a database of area and energy-per-access costs that is created by generating and measuring a large variety of memory structures with different parameters using, for example, a memory compiler in a particular technology.

(2) An *Arithmetic Model*, which supports modeling of MACs with different bit-widths. Such a model may be based on, for example, a database of area and energy per access costs created by synthesizing and measuring various multiplier and adder designs for different bit-widths in a given technology. For bit-widths not in the database, the model may scale energy per access using analytical methods, e.g., quadratically (for multipliers) or linearly (for adders).

(3) A *Wire/Network Model*, which supports the modeling of on-chip networks with different topologies and bit-widths. To model the wire energy consumed for a network hop, area estimates can be used to determine the wire distance between the two ends of the hop. This distance can be multiplied with a fJ/bit/mm value measured in a given technology to determine the per-hop energy.

6.4   BUILDING AN AUTOMATED MAPPER

A *mapspace* is the set of all legal mappings of a workload onto an architecture. The structured mapping representations described in Section 6.2 enable us to programmatically enumerate this set. For example, in a loop nest representation, all mappings in a mapspace have the same number of tiling levels, but differ in (a) the values assigned to the loop bounds at each level and (b) the permutation of loops within each level. To construct the mapspace, we must enumerate all possible factorizations of each workload dimension across levels, all possible permutations of loops within a level, and all level bypassing alternatives. The Cartesian product of these choices gives us an *unconstrained* mapspace, which can be quite large due to combinatorial explosion, e.g., for a CONV2D layer on a 4-tiling-level architecture the size is $(7!)^4 \times (2^4)^3 \times$ size of the Cartesian product of the co-factor sets for each of the 7 loop bounds. While there are ways to prune this space—e.g., permutations do not matter for the innermost tiling level—the space is still large.

However, most architectures do not have enough flexibility to support such an unconstrained mapspace. Mapper infrastructures allow users to capture these hardware capabilities as *constraints* [88]. These user-specified hardware constraints are accommodated into the mapspace, shrinking its size.

With the mapspace constructed in a systematic fashion, a search routine can search through the space for an optimal mapping based on user-provided optimization criteria. A mapping sampled from the mapspace can be evaluated using the model (Section 6.3), and the derived statistics can be used to compare mappings based on the optimization criteria. Various search heuristics have been used in prior work, including genetic algorithms [92], beam search [93, 94], parallel random sampling [88], reinforcement learning [95], parallel simulated annealing [96], and exhaustive search [97].

6.5   SUMMARY

In this chapter, we first discussed the need to separate the notion of an architecture design space of DNN accelerators from the mapping space for a specific workload on an accelerator. We described two alternative approaches to representing a mapping. We discussed how contemporary DNN accelerator modeling infrastructures model the execution of a mapping on a specific architecture design in a fast and precise manner. Finally, we described how a mapper can leverage a structured mapping representation and a fast analytical model to automatically search the mapping space for an optimal mapping.

CHAPTER 7

Orchestrating Compressed-Sparse Data

DNN operations often involve a considerable amount of zeros in activations and filter weights because of nonlinear activation functions, weight pruning, and so on. Since a zero operand, either filter or activation, results in a zero partial sum (or, multiple of them), the corresponding partial sums do not contribute to the output activation. Therefore, we can skip computation of such redundant computations and corresponding data communication to save energy and reduce latency. However, compressed sparse data lead to irregular dataflow, and it requires various meta data to track the data indices in the original data space. In particular, the meta data processing costs can be significant depending on the sparsity ratio and compression format, orchestrating compressed sparse data is a challenging problem. In this chapter, we provide background about the sparsity in DNNs and discuss state-of-the-art dataflow styles on compressed sparse data.

7.1 OVERVIEW

Research has shown that many contemporary neural networks have significant redundancy and can be *pruned* dramatically during training without significantly affecting accuracy [98]. The number of weights that can be eliminated varies widely across the layers but typically ranges from 20–80% [98, 99] for contemporary networks. Eliminating weights results in a network with a substantial number of zero values, which can potentially reduce the computational requirements of inference.

The inference computation also offers a further optimization opportunity, as many networks employ as their nonlinear operator the ReLU (rectified linear unit) function which clamps all negative activation values to zero. The activations are the output values of an individual layer that are passed as inputs to the next layer. For typical data sets, 50–70% of the activations are clamped to zero. Since the multiplication of weights and activations is the key computation for inference, the combination of these two factors can reduce the amount of computation required by over an order of magnitude. Additional benefits can be achieved by a compressed encoding for zero weights and activations, thus allowing more to fit in on-chip RAM and eliminating energy-costly DRAM accesses.

A number of contemporary inference accelerators take advantage of one or more of the above forms of sparsity. For example, the Eyeriss [20, 64] accelerator saves computation energy for zero-valued activations and compresses weights and activations stored in DRAM. Other approaches use either a compressed encoding of activations [100] or compressed weights [101] in parts of their dataflow to reduce data transfer bandwidth and save time for computations of some multiplications with a zero operand. The SCNN [27] accelerator exploits both weight and activation sparsity to improve the performance and power of DNNs. SCNN employs both an algorithmic dataflow that eliminates all multiplications with a zero and a compressed representation of both weights and activations through almost the entire computation.

Compressed sparse architectures have typically been shown to provide a factor of 2-3× speedup and energy reduction relative to a comparably provisioned dense DNN accelerator [27].

7.2   SPARSITY IN DNNS

Sparsity in a DNN layer is defined as the fraction of zeros in the layer's weight and input activation matrices. The primary technique for creating weight sparsity is to prune the network during training. Han et al. developed a pruning algorithm that operates in two phases [98]. First, any weight with an absolute value that is close to zero (e.g., below a defined threshold) is set to zero. This process has the effect of removing weights from the filters, sometimes even forcing an output activation to always be zero. Second, the remaining network is retrained, to regain the accuracy lost through naïve pruning. The result is a smaller network with accuracy extremely close to the original network. The process can be iteratively repeated to reduce network size while maintaining accuracy.

Activation sparsity occurs dynamically during inference and is highly dependent on the data being processed. Specifically, the rectified linear unit (ReLU) function that is commonly used as the nonlinear operator in DNNs forces all negatively valued activations to be clamped to zero. After completing computation of a convolutional layer, a ReLU function is applied pointwise to each element in the output activation matrices before the data is passed to the next layer.

Figure 7.1 shows the weight and activation *density* (fraction of non-zeros, i.e., the complement of sparsity) of a subset of representative layers of three contemporary networks (weights were pruned using the algorithm in [98]). The density fractions are represented on the left-hand Y-axes of the charts. The data shows that weight density varies across both layers and networks, reaching a minimum of 30% for some of the GoogLeNet layers. Activation density also varies, with density typically being higher in early layers. Activation density can be as low as 30% as well. The triangles show the ideal number of multiplies that could be achieved if all multiplies with a zero operand are eliminated. This is calculated by by taking the product of the weight and activation densities on a per-layer basis.

7.3   STRUCTURED VS. UNSTRUCTURED SPARSITY

Distribution of non-zero values in activation and pruned-weight tensors in general does not exhibit any regular patterns. This natural form of sparsity is known as *unstructured sparsity*. However, pruning of weights can be performed in such a way as to enforce the zeros to appear in

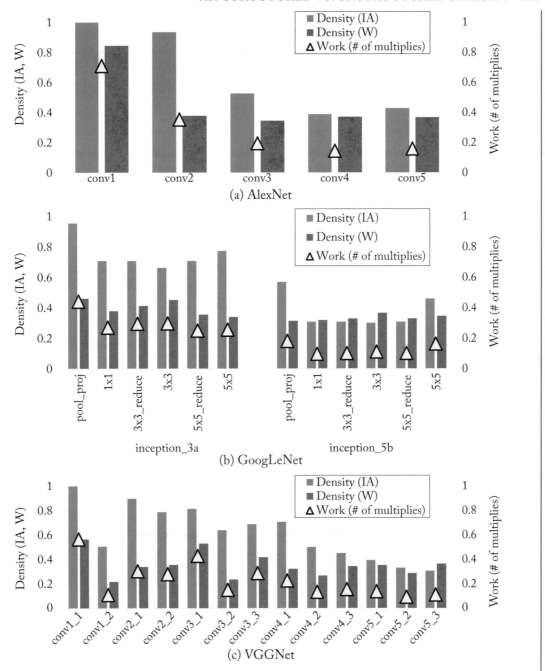

Figure 7.1: Input activation and weight density and the reduction in the amount of work achievable by exploiting sparsity.

certain constrained patterns. This form of sparsity is known as *structured* sparsity. Forcing a structure on the sparsity pattern allows for simpler dataflows and microarchitectures. However, for a given compressed model size, unstructured-sparse models produce higher accuracy. Furthermore, activation sparsity has not been shown to be constrained into structured patterns. Therefore, we focus our attention on *unstructured* sparsity in the remainder of this chapter.

7.4 EXPLOITING SPARSITY

Since multiplication by zero just results in a zero, it should require no work. Thus, typical layers can reduce work by a factor of four, and can reach as high as a factor of ten. In addition, those zero products will contribute nothing to the partial sum it is part of, so the addition is unnecessary as well. Furthermore, data with many zeros can be represented in a compressed form. Together these characteristics provide a number of opportunities for optimization.

- **Compressing data:** Encoding the sparse weights and/or activations provides an architecture an opportunity to reduce the amount of data that must be moved throughout the memory hierarchy. It also reduces the data footprint, which allows larger matrices to be held in a storage structure of a given size.

- **Eliminating computation:** For multiplications that have a zero weight and/or activation operand, the operation can be data gated, or the operands might never be sent to the multiplier. This optimization can save energy consumption or both time and energy consumption, respectively.

Eyeriss [20] gates the multiplier when it sees an input activation of zero [69]. Because the sparsity of weights is not significant for non-pruned networks, Eyeriss opted not to gate the multiplier on zero weights. This gating approach will save energy, but not execution time.

Another approach to reducing energy is to reduce data transfer costs when the data is sparse. Eyeriss uses a run length encoding scheme when transferring activations to and from DRAM. This approach saves energy (and time) by reducing the number of DRAM accesses. However, because the data is kept in an expanded form in the on-chip memory hierarchy, such architectures cannot completely eliminate energy on the data transfers from one internal buffer to another internal buffer or to the multipliers. Eyeriss does, however, save the energy for accessing the weight buffer when the associated activation is zero.

Cnvlutin [100] is more aggressive—the architecture moves and stages sparse activations in compressed form and skips computation cycles for zero-valued activations to improve both performance and energy efficiency. Both Eyeriss and Cnvlutin are also able to partially elide inner-buffer accesses for *weights* if those weights were only to be multiplied with a zero-valued activation.

The Cambricon-X [101] architecture exploits weight sparsity by keeping only non-zero weights in its internal buffers [101] in compressed form. The architecture is able to skip computation cycles for zero-valued weights. However, it does not compress activation data between

DRAM and the accelerator. Nor does it keep activations in a compressed form in the internal buffers, except in the queues directly delivering activations to the multipliers. It suffers from wasted computation cycles when the non-zero weight is to be multiplied with zero-valued activations.

The EIE accelerator uses a compressed representation of both activations and weights, and only delivers non-zero operands to the multipliers [28]. EIE is specifically designed for the fully connected layers of a DNN model.

The SCNN architecture [27] exploits sparseness in both activations and pruned weights to eliminate as many computation cycles and data movement and storage operations as possible. SCNN employs a compressed encoding of both sparse weights and activations so that only non-zero data values are retrieved from DRAM, and also maintains that compressed representation in all the on-die buffers. This approach saves time and energy on all data transfers and allows the chip hold larger models for a given amount of internal storage. For many networks, SCNN can typically keep all of the activations on-chip between layers, further improving performance and energy-efficiency.

7.5   SPARSE DATAFLOWS

In addition to the different approaches of exploiting sparsity, various architectures employ distinct dataflows [20] to execute a sparse convolutional layer. The most relevant distinction among these architectures' dataflows is how the innermost computation datapath exploits *spatial* reuse and sparsity patterns. For example, Eyeriss uses a row-stationary dataflow, multicasting weights and activations across multiple scalar PEs, with each PE independently performing zero-activation detection. Cnvlutin multiplies a single scalar non-zero activation across a vector of weights (organized by output-channel), and then reduces these output vectors across different input-channels. Cambricon-X fetches activation vectors across input-channels based on non-zero weight vectors and computes their dot product, including unnecessary work for zero-valued elements of the activation vector. SCNN employs a *Cartesian product* dataflow that exploits both weight and activation reuse while delivering only non-zero weights and activations to the multipliers. This dataflow performs an all-to-all multiply of non-zero weight and activation vector elements that can avoid any arithmetic based on zero-valued operands and achieve full multiplier utilization in steady-state. Table 7.1 summarizes all the attributes of these accelerators.

To develop a better understanding of the complexities involved with orchestrating sparse data, we now take a deep dive into the dataflow employed by the SCNN [27] architecture. While the inner core of the dataflow is based on a spatial Cartesian product, the complete dataflow requires a deep nested loop structure, mapped both spatially and temporally across multiple processing elements. The full dataflow is called *PlanarTiled-InputStationary-CartesianProduct-Sparse*, or PT-IS-CP-Sparse. PT-IS-CP-Sparse enables reuse patterns that exploit the characteristics of sparse weights and activations. This section first presents an equivalent dense dataflow

Table 7.1: Qualitative comparison of sparse CNN accelerators. The letters "A" and "W" denote whether the architecture in that row exploits sparsity in Input Activations and/or Weights, respectively, by performing the optimization in that column.

Architecture	Gate MACC	Skip MACC	Skip Buffer/ DRAM Access	Inner Spatial Dataflow
Eyeriss [20]	A	–	A	Row Stationary
Cnvlutin [100]	A	A	A	Vector Scalar + Reduction
Cambricon-X [101]	W	W	W	Dot Product
SCNN [27]	A+W	A+W	A+W	Cartesian Product

(PT-IS-CP-Dense) to explain the decomposition of the computations and then adds the specific features for PT-IS-CP-Sparse.

7.5.1 THE PT-IS-CP-DENSE DATAFLOW

Single-multiplier temporal dataflow. The *IS* term in PT-IS-CP-Dense describes the *temporal* component of the dataflow. First, consider the operation of a scalar PE with a single multiply-accumulate unit. The dataflow employs an *input-stationary* (IS) computation order in which an input activation is held stationary at the computation units as it is multiplied by all of the filter weights needed to make all of its contributions to each of the K output channels (a $K \times R \times S$ sub-volume). Thus, each input activation will contribute to a volume of $K \times W \times H$ output activations. This order maximizes the reuse of the input activations, while paying a cost to stream the weights to the computation units. More precisely, the register in which the stationary input is held over $K \times R \times S$ iterations serves as an inner buffer that filters accesses to the larger input buffer. Accommodating multiple input channels (C) adds an additional outer loop and results in the loop nest $C \to W \to H \to K \to R \to S$.

The PT-IS-CP-Dense dataflow requires input buffers for weights and input activations, and an accumulator buffer to store the *partial sums* of the output activations. The accumulator buffer must perform a read-add-write operation for every access to a previously-written index.

Unfortunately, the stationarity of input activations comes at the cost of more streaming accesses to the weights and to the partial sums in the accumulator buffer. Blocking the weights and partial sums in the output channel (K) dimension can increase reuse of these data structures and improve energy efficiency. The dataflow therefore factors the K output channels into K/K_c *output-channel groups* of size K_c, and only store weights and outputs for a single output-channel group at a time inside the weight and accumulator buffers. Thus, the sub-volumes that are housed in buffers at the computation unit are:

- Weights: $C \times K_c \times R \times S$

- Inputs: $C \times W \times H$

- Partial Sums: $K_c \times W \times H$

An outer loop over all the K/K_c output-channel groups results in the complete loop nest $K/K_c \to C \to W \to H \to K_c \to R \to S$. Each iteration of this outer loop will require the weight buffer to be refilled and the accumulator buffer to be drained and cleared, while the contents of the input buffer will be fully reused because the same input activations are used across all output channels.

Intra-PE parallelism. The *CP* term in PT-IS-CP-Dense describes how parallelism of many multipliers within a PE can be exploited while maximizing *spatial* reuse. A vector of F filter-weights fetched from the weight buffer and a vector of I inputs fetched from the input activation buffer are delivered to an array of $F \times I$ multipliers to compute a full Cartesian product (CP) of output partial-sums. This *all-to-all* operation has two useful properties. First, each fetched weight is reused (via wire-based multicast) over all I activations; each activation is reused over all F weights. Second, each product yields a useful partial sum such that no extraneous fetches or computations are performed. PT-IS-CP-Sparse will exploit these same properties to make computation efficient on compressed-sparse weights and input activations.

The multiplier outputs are sent to the accumulation unit, which updates the partial sums of the output activation. Each multiplier output is accumulated with a partial sum at the matching output coordinates in the output activation space. These coordinates are computed in parallel with the multiplications. The accumulation unit must employ at least $F \times I$ adders to match the throughput of the multipliers.

Inter-PE parallelism. Finally, the *PT* term in PT-IS-CP-Dense describes how to scale beyond the practical limits of multiplier count and buffer sizes within a PE. The dataflow employs a spatial tiling strategy to spread the work across an array of PEs so that each PE can operate independently. The $W \times H$ element activation plane is partitioned into smaller $W_t \times H_t$ element *planar tiles* (PT) that are distributed across the PEs. Each tile extends fully into the input-channel dimension C, resulting in an input-activation volume of $C \times W_t \times H_t$ assigned to each PE. Weights are broadcast to the PEs, and each PE operates on its own subset of the input and output activation space.

Unfortunately, strictly partitioning both input and output activations into $W_t \times H_t$ tiles does not work because the sliding-window nature of the convolution operation introduces cross-tile dependencies at tile edges. These data *halos* (Section 2.2, [74]) can be resolved in one of two ways.

- **Input halos:** The input buffers at each PE are sized to be slightly larger than $C \times W_t \times H_t$ to accommodate the halos. These halo input values are replicated across adjacent PEs, but outputs are strictly private to each PE. Replicated input values can be multicast when they are being fetched into the buffers.

```
          BUFFER wt_buf[C][Kc*R*S/F][F];
          BUFFER in_buf[C][Wt*Ht/I][I];
          BUFFER acc_buf[Kc][Wt+R-1][Ht+S-1];
          BUFFER out_buf[K/Kc][Kc*Wt*Ht];
(A)  for k' = 0 to K/Kc-1
     {
         for c = 0 to C-1
           for a = 0 to (Wt*Ht/I)-1
           {
(B)          in[0:I-1] = in_buf[c][a][0:I-1];
(C)          for w = 0 to (Kc*R*S/F)-1
             {
(D)            wt[0:F-1] = wt_buf[c][w][0:F-1];
(E)            parallel_for (i = 0 to I-1) x (f = 0 to F-1)
               {
                 k = Kcoord(w,f);
                 x = Xcoord(a,i,w,f);
                 y = Ycoord(a,i,w,f);
(F)              acc_buf[k][x][y] += in[i]*wt[f];
               }
             }
           }
         out_buf[k'][0:Kc*Wt*Ht-1] =
           acc_buf[0:Kc-1][0:Wt-1][0:Ht-1];
     }
```

Figure 7.2: PT-IS-CP-Dense dataflow, single-PE loop nest.

- **Output halos:** The accumulation buffers at each PE are sized to be slightly larger than $K_c \times W_t \times H_t$ to accommodate the halos. The halos now contain incomplete partial sums that must be communicated to neighbor PEs for accumulation, which occurs at the end of computing each output-channel group.

Figure 7.2 shows pseudo-code for a single PE's loop nest in the PT-IS-CP-Dense dataflow, including blocking in the K dimension (A,C), fetching vectors of input activations and weights (B,D), and computing the Cartesian product in parallel (E,F). X and Y coordinates for the accumulation buffer that are either negative or greater than $W_t - 1$ and $H_t - 1$ correspond to the locations of incomplete partial sums in the halo regions. Communication of these halos to neighboring PEs is not shown in the figure. The *Kcoord*(), *Xcoord*(), and *Ycoord*() functions compute the k, x, and y coordinates of the uncompressed output volume using a delinearization of the temporal loop indices a and w, the spatial loop indices i and f, and the known filter width and height.

7.5.2   PT-IS-CP-SPARSE DATAFLOW

PT-IS-CP-Sparse is a natural extension of PT-IS-CP-Dense that exploits sparsity in the weights and input activations. The dataflow is specifically designed to operate on compressed-sparse encodings of the weights and input activations and produces a compressed-sparse encoding of the output activations. At a CNN layer boundary, the output activations of the previous layer become the input activations of the next layer. The specific compression format used [28, 100, 101] is orthogonal to the sparse architecture itself. The key feature is that decoding a sparse format ultimately yields a non-zero data value and an index indicating the coordinates of the value in the weight or input activation matrices.

To facilitate easier decoding of the compressed-sparse blocks, weights are grouped into compressed-sparse blocks at the granularity of an output-channel group, with $K_c \times R \times S$ weights encoded into one compressed block. Likewise, input activations are encoded at the granularity of input channels, with a block of $W_t \times H_t$ encoded into one compressed block. At each access, the weight buffer delivers a vector of F **non-zero** filter weights along with each of their coordinates within the $K_c \times R \times S$ region. Similarly, the input buffer delivers a vector of I **non-zero** input activations along with each of their coordinates within the $W_t \times H_t$ region. Similar to the dense dataflow, the multiplier array computes the full cross-product of $F \times I$ partial sum outputs, with no extraneous computations. Unlike a dense architecture, output coordinates are not derived from loop indices in a state machine but from the coordinates of non-zero values embedded in the compressed format.

Even though calculating output coordinates is trivial, the multiplier outputs are not typically contiguous as they are in PT-IS-CP-Dense. Thus, the $F \times I$ multiplier outputs must be scattered to discontiguous addresses within the $K_c \times W_t \times H_t$ output range. Because any value in the output range can be non-zero, the accumulation buffer must be kept in an uncompressed format. In fact, output activations will probabilistically have high density even with a very low density of weights and input activations, until they pass through a ReLU operation.

To accommodate the needs of accumulation of sparse partial sums, the monolithic $K_c \times W_t \times H_t$ accumulation buffer from the PT-IS-CP-Dense dataflow is split into a distributed array of smaller accumulation buffers, which are accessed via a scatter network that can be implemented as a crossbar switch. The scatter network routes an array of $F \times I$ partial sums to an array of A accumulator banks based on the output index associated with each partial sum. Taken together, the complete accumulator array still maps the same $K_c \times W_t \times H_t$ address range, although the address space is now split across a distributed set of banks. PT-IS-CP-Sparse can be implemented via small adjustments of Figure 7.2. Instead of a dense vector fetches, (B) and (D) fetch the compressed sparse input activations and weights, respectively. In addition, the coordinates of the non-zero values in the compressed-sparse form of these data structures must be fetched from their respective buffers (not shown). Then the accumulator buffer (F) must be indexed with the computed output coordinates from the sparse weights and activations. Finally, when the computation for the output-channel group has been completed, the accumulator buffer is drained and compressed into the output buffer.

A microarchitecture implementing the PT-IS-CP-Sparse dataflow is described in detail in [27].

7.6   COSTS AND BENEFITS

At this point, it is hopefully evident that while sparsity is abundant in both activations and pruned weights, exploiting this sparsity often requires more complex dataflows and microarchitectural designs. This complexity manifests itself in terms of additional performance and energy overheads. Fortunately, nearly all studies on exploiting sparsity demonstrate that for typical sparsity patterns observed on contemporary networks, the benefits almost universally outweigh the costs. For example, the SCNN [27] architecture achieves a speedup of 2.7× and an energy-efficiency improvement of 2.3× over a dense accelerator.

A reasonable question to ask at this stage is—how would these architectures behave if the sparsity (perhaps, of a future network) is lower than the patterns we are measuring today? Would the overheads begin to outweigh the benefits? More precisely, we can attempt to find a *crossover point* in terms of density above which a sparse architecture is not worth the complexity. In [27], the authors artificially sweep the density of a synthetic workload to establish this crossover point for a variety of architectures and show that it ranges from 60–100% density (i.e., fraction of non-zeroes) depending on the architecture and the approach used to exploit sparsity.

Understanding energy and performance overheads. The relatively lower energy efficiency of sparse architectures running dense workloads arises from the interplay of multiple attributes of their dataflows. However, one dominant factor is that they cannot exploit the sheer efficiency of a spatial reduction tree. This is explored in detail in [102].

Performance overheads are arguably more interesting, and are primarily caused by the *irregularity* of the compressed-sparse data. This leads to two sources of inefficiency. First, the architecture can suffer from *fragmentation* when there isn't enough useful work to fully populate vectorized arithmetic units (e.g., in the last temporal iteration of a loop). Second, uneven partitioning of work across coarse-grained units of computation (such as PEs) may lead to *load imbalance*. A quantitative analysis of these effects is covered in [27]. Load imbalance and fragmentation are likely to be universally-visible effects across all compressed-sparse architectures, although the precise extent to which these effects affect performance depends on the dataflow and microarchitectural attributes of the architecture.

7.7   SUMMARY

Much of this chapter focused on sophisticated techniques to exploit unstructured sparsity for performance and energy-efficiency gains. As we have seen, these techniques require relatively complex dataflows and microarchitectures to extract these benefits, which imposes certain overheads for dense workloads but yield strong dividends or sufficiently sparse workloads. In contrast, we have seen that simple optimizations such as zero-gating multiplier operands and compressing data between storage levels are remarkably effective at achieving good energy efficiency on moderately sparse workloads, although they cannot improve performance. We also noted in Section 7.3 that *structured* sparsity can be imposed on pruned weights, which allows for simpler dense-like architectures to obtain some of the performance and energy-efficiency benefits of architectures designed for unstructured sparsity. This exposes an interesting tradeoff space that architects must navigate, with careful consideration of the properties workloads that are expected to run on an architecture under design.

CHAPTER 8

Conclusions

In this book, we systematically described various data orchestration mechanisms employed in DNN accelerators. We defined and delved into dataflows—the mechanism for extracting data reuse from the algorithm. Data reuse in turn leads to reduced data movement, enhancing performance and energy-efficiency. Next, we discussed how dataflows can be realized via efficient custom memory hierarchy and on-chip network implementations. We then discussed the design-flow involved in DNN accelerator design via case studies, and also highlighted the role of microarchitectural models and mappers during the design and deployment process. Finally, we gave a glimpse of the challenges and opportunities in designing accelerators for sparse and compressed DNNs.

In this chapter, we list some compelling research opportunities in data orchestration for inference ASIC accelerators (the focus of this book). We also point to data orchestration challenges in alternate inference platforms (Analog, FPGA, and GPUs) that were beyond the scope of this book, but are nonetheless areas of active research in the community. We also briefly discuss opportunities in designing accelerators for training.

8.1 RESEARCH OPPORTUNITIES IN DATA ORCHESTRATION

8.1.1 DATAFLOW-HARDWARE CO-DESIGN AND FLEXIBILITY

In this book, we discussed how a specific dataflow strategy can be realized with various hardware implementations for the buffers and interconnect. Each of these implementations can provide levels of performance and flexibility and come at different area or power costs. The design-space of the cross-product of dataflows and hardware architectures is extremely large and there are multiple research opportunities for co-design. Today's accelerator implementations barely scratch the surface of such co-design opportunities.

Similarly, another axis of research relates to the amount of flexibility an accelerator should support across dataflows. On one end could be a hardware architecture that is extremely flexible but comes at the cost of highly configurable buffers and interconnects; one the other end could be an architecture that is inflexible but cheap to implement. It is an open question on whether the area within the accelerator should be devoted toward more complex structures that extract more reuse and provide higher performance vs. simple but underutilized structures. The answers to this question also touch upon the basic philosophy behind accelerator design—how much loss in efficiency due to programmability is desirable in accelerators.

At first glance it may seem strange to mention programmability in the context of accelerators at all. Isn't the entire goal to offload computation from our programmable platforms? The reason field-programmability is important is that the field of machine learning is changing very rapidly, and architecture-to-silicon design process for accelerators still takes roughly two years. Currently, there is a very real chance that any accelerator deployed could be obsolete before it reaches the market, because an algorithm change will have achieved the same (or more) benefits. Field-programmability is a hedge against this risk, and allows a certain amount of "future-proofing" to be built in to the accelerator. Unfortunately, there is currently no way to quantify the benefits of programmability, nor to objectively make statements such as "accelerator X is twice as programmable as accelerator Y." As such, research effort is needed to deeply characterize the cost/benefit tradeoff of programmable accelerator features.

Finally, a third axis of research is in the field of finding an optimal mapping of a workload on a specific architecture with a given degree of flexibility. The mapping formulation we covered in this book is sufficient to cover regular workloads such as specific DNN layers. However, even for these workloads the mapspaces for flexible architectures are so large that search heuristics often struggle to find an optimal solution. Expanding the scope of workloads to cover complete DNNs will exacerbate this problem. Mapspace formulation and efficient heuristics to search them therefore remain active and important research problems. In fact, the design space search for accelerators and mapspace search for a given accelerator together form a very challenging co-optimization problem.

8.1.2 QUANTIZATION

When looking to optimize data movement, one natural line of research is to investigate whether we can reduce the amount of data transferred. Weight *quantization* is an approach where the total number of unique values among the filters is minimized, usually as part of the training process [98, 103, 104]. These remaining values are encoded using compression techniques.

Accelerators can leverage quantization by either expanding the values before multiplication [105], or by memo-izing the multiplication operation itself over the existing weights [106] (assuming the input activation bitwidth is low enough to make this profitable). While this can drastically reduce data transferred, it can have a significant effect on inference accuracy, and more research is needed to justify the savings over non-quantized networks.

8.1.3 SPARSITY

One of the most promising areas of optimization for DNN accelerators is in the area of sparsity. While early DNN accelerators prototypes, including most of the ones discussed throughout this book, targeted dense and well-structured DNNs, emerging DNNs are highly irregular and sparse as their architecture is driven by automated neural-architecture search (NAS) and explicit sparsification during training. Moreover, as AI becomes pervasive, DNN models deployed for applications in computational science and engineering are expected to exhibit extremely high levels of sparsity, often more than 99%. Furthermore, research has shown that sparsity can go up further with less accuracy loss when it is included as an explicit goal in the training process [107]. At these levels of sparsity using a compressed data format becomes profitable.

When considering compression formats, a lot of the reuse analysis, buffer management, and data distribution and collection approaches discussed in the book need to be expanded or reconsidered entirely. For example, rather than assuming that we know a tile's shape, a mapper may need probabilistic notions of tile occupancy. However, sparse data is usually not uniformly distributed, and so this becomes quite challenging as the mapper needs to reason about both worst case and average case.

There have been some recent research efforts into reducing meta-data access costs for such workloads [76, 108–110]. As we begin to approach the limits of reuse exploitation for dense workloads, we expect to see many accelerator designs focusing on sparsity, and being able to run a combination of dense and sparse workloads efficiently.

8.1.4   ALGEBRAIC RE-ASSOCIATION AND CO-DESIGN

Exploiting sparsity is based on the algebraic property that $\forall x . x \times 0 = 0$ but there are other algebraic properties that we can exploit to reduce computation. For example in the well-known Winograd approach [111, 112] the fundamental neuron's sum-of-products is refactored into a common subset, and neuron-specific residuals. While this results in fewer total multiplications, it also increases irregularity.

One way to minimize the negative impact of irregularity is to co-design the re-association and the accelerator itself, as in the UCNN accelerator [113]. UCNN de-duplicates the weight values and then performs multiplication across all common operands. The zero operands are dropped—due to the above algebraic property—so UCNN's approach can be seen as a generalization of sparsity. The approach's benefits increase when combined with weight quantization, but it is an open question whether quantization is broadly acceptable to the field.

8.1.5   SERIAL ARITHMETIC

Beyond algebraic properties, some accelerators look to exploit the fundamental arithmetic characteristics of the MAC operation. For example, in most multiplications, the high bits are likely to be zero (excluding sign extension). Therefore, it is worth exploring whether spatially dedicating circuits exclusively to these high-bits is an efficient use of area. There has been active research into whether it may be better to temporally re-use smaller-width multipliers [114], supplemented with control FSMs. The most extreme form of this is fully bit-serial arithmetic [115]. Like sparsity, these approaches impose challenges of irregular and unpredictable processing time. Research is required to adapt traditional data orchestration techniques to these platforms.

8.1.6   SCALING ACCELERATOR SYSTEMS

Advances in the machine learning community has led to development of new networks which significantly improve the accuracy on tasks like recommendation and language modeling [12, 72]. However, these new improved models are larger and therefore demand massive computing power. This brings us to one of the pertinent recent topics in accelerator design: building accelerators at scale.

There are two major ways in which the accelerators can be scaled. The first method is to simply make larger accelerators, with more computing units (*aka scale-up*). The first version of Google's TPU [25] is an illustrative example of this approach. As we have seen in the previous chapters, this design has a massive 256×256 systolic array of MAC units. A more impressive and perhaps radical design example is the recent Wafer Scale Processor, built by Cerebras system, which has an unprecedented area footprint of 46,225 mm^2 containing more than 1.2 trillion transistors [116]. The motivation behind building such a large chip is to keep the memory near the compute, and therefore attain energy and performance degradation arising from frequent off-chip accesses. Unlike the accelerators we discussed throughout the book, the prime workload for such architectures is DNN training. For inference workloads however, utilizing such a large monolithic block of compute often leads to under-utilization, especially for irregular and sparse DNNs.

This brings us to our second method of the scaling, where multiple smaller accelerator units are used together to solve a given problem (a.k.a. scale-out). One recent example of such a system is SIMBA [61] from NVIDIA. SIMBA comprises of 36 chiplet modules, each of which is a standalone accelerator, connected with each other and the external peripherals via on-chip interposer. Yet another example is the Tetris system [74] proposed by researchers at Stanford university. Tetris uses a collection of Eyeriss [20] like accelerators, each with a dedicated 3D stacked memory unit. The problem with such distributed systems is the immediate loss of spatial reuse as communication among the distinct computation units is much expensive than a monolithic system.

Finding the right mapping and data orchestration strategy for such systems is of paramount importance to enable efficient computation. Furthermore, it is also not intuitive to determine whether a system should be made purely monolithic, or purely distributed by employing computation units as soon as possible, or alternatively taking the middle ground by building appropriately sized processors which are then employed as units in a distributed setting. Further research in this direction therefore is timely and has tremendous academic and economic merits.

8.2 DATA ORCHESTRATION IN ALTERNATE PLATFORMS

8.2.1 ANALOG

An emerging class of DNN inference accelerators is based on analog computing—leveraging current summation within dense analog arrays to perform large vector-matrix computations extremely efficiently, allowing both high-speed and high-energy efficiency [117]. These architectures are called Compute in Memory (CIM). These analog arrays can be built using various memory technologies: SRAM, Resistive RAM (RRAM), phase change memory (PCM), Spin-transfer Torque Magnetic RAM (STT-MRAM), and so on. The key idea in CIMs is to encode the weights within the memory structures, and feed in the input activations as currents. These dense analog arrays have shown higher energy-efficiency than performing the same computation via digital MACs. However, these analog accelerators face four key challenges before they can get to the mainstream. First, programming weights into CIMs is a highly time and energy consuming process—thus CIM arrays need to inherently operate using a weight-stationary dataflow. In fact, the key benefit of the arrays comes only when the entire model is stationary throughout the operation. This limits their applicability to smaller DNN models. Second, analog current summation is inherently noisy. This can lead to potential accuracy losses during inference. To address this, typically DNN models that will be run on such analog accelerators need to be trained keeping this error tolerance in mind. Third, the Analog-to-Digital converters at the ends of the array often consume high energy and power, diminishing some of the overall energy-efficiency benefits. And finally, CIMs are highly efficient for dense vector-matrix multiplications, but not for sparse DNNs.

8.2.2 FPGA

FPGAs have emerged as a popular choice for DNN inference, as their reconfigurable substrate helps tailor the datapath for the DNN model. Most of the DNN accelerators discussed throughout the book can all be mapped as overlay architectures on FPGAs. For high performance, the MAC units are typically mapped over the DSPs and the scratchpad memories over the BRAMs. The distribution and collection NoCs get mapped over the configurable logic.

One key optimization that FPGAs offer that ASIC accelerators typically cannot support is bit-level configurability. FPGAs can support lower bit-precision (down to 1-bit precision, e.g., for binary neural networks), and/or separate bit-precision for compute within different layers. This enables FPGAs to run quantized DNNs much more efficiently than their ASIC counterparts.

The key shortcoming of FPGAs remains their operating frequency, which is much lower than the ASIC accelerators that have hardened datapaths. Thus, getting performance out of FPGA-based accelerators requires a FPGA architecture-aware mapping flow—since the relative area (and power) cost of compute, memory, and wires on a silicon tapeout is different from that on the FPGA, for e.g., building many of the complex compute, buffer, or interconnect components described in this book using configurable LUTs, while providing flexibility, is much more expensive than their hardened ASIC counterparts. Thus, customized libraries like Xilinx xDNN [118] implement store-and-forward distribution and collection over direct hardened links between the DSPs, to avoid going into the configurable logic to boost up operating frequency.

The rise of AI and DNNs has led to FPGA vendors actively changing the underlying FPGA architecture to make them competitive to their ASIC counterparts. For example, Xilinx's Versal AI engine [119] houses a spatial array of hardened AI cores, not unlike the DNN accelerators discussed throughout this book, enabling workloads to get mapped between the AI cores and the conventional configurable logic.

8.2.3   GPU

GPUs were the hardware platform that propelled the Deep Learning revolution in the first place since they house hundreds to thousands of ALUs which can be leveraged to accelerate matrix multiplications. From a 1000-foot view, GPUs do not look too different from the DNN accelerators described in this book: they come with arrays of PEs, scratchpad memories, and interconnects. A few fundamental differences exist however. First, the PEs in the GPU are arranged as SIMD lanes, and do not provide the ability for direct communication for spatial reuse. All communication in the GPUs is via registers and memory. Second, GPUs are fundamentally programmed as control-flow, and cannot exploit dataflow like DNN accelerators. Third, the GPU memory hierarchy is built for general programs, unlike DNN accelerator hierarchies optimized for DNN operands. Fourth, GPU datapaths are typically provisioned to maximize higer-precision floating point throughput, while the PEs in DNN inference accelerators tend to use lower-precision fixed-point datapaths. Thus, GPUs remain the most popular choice for DNN training, while custom accelerators continue to be built for inference (especially for mass deployment on energy-constrained edge devices).

There has been interesting trend, however, in recent GPU architectures, to answer the rise in specialized DNN accelerators. NVIDIA's GPUs, since the Volta [120] product line from 2017, offer "Tensor Cores" within each Streaming Multiprocessor (SM). Tensor cores are tiny engines optimized for running 4×4 GEMMs. In other words, each SM within NVIDIA GPUs now comes with a specialized acceleration block for DNNs, which incorporates several of the attributes of dedicated DNN accelerators such as optimized dataflows (including spatial reuse exploitation) and lower-precision data formats and operators favored for inference operations. More recently, NVIDIA's Ampere [121] GPUs released in 2020 provided support to handle structured sparsity within the tensor cores. Thus, GPUs are slowly but steadily incorporating some of the principles for DNN acceleration discussed in this book.

8.3 TRAINING ACCELERATORS

In this book, we focused on the design-flow for inference accelerators deployed across various form factors. Specialized DNN accelerators like the Google TPU [25], however, also support training. We briefly discuss some of the design considerations here, highlighting some fundamental differences that makes the training workloads stand apart from inference ones.

First training involves computation of weight and input gradients in addition to forward pass which is performed in inference. Gradient computation and weight updates demonstrate *Weak Scaling*, which means the time to convergence does not strictly decrease with the increase in the speed at which computation is performed. This is due to the fact that convergence depends on the amount of data on which weight updates are performed. Achieving fast convergence on training therefore requires careful design of both the compute units and the communication framework.

Second, computations in inference can be further optimized by using reduced precision operands, which help both reduce the cost of computation and memory footprint. During training, however, such optimizations are not possible. Training accelerators need to support floating point operation.

Third, in case of training, the intermediate activations are needed to be stored along with the input and weight matrices, which are then used in the backpropagation and weight update stages. For modern networks the sizes of all the matrices are usually in the order of tens of GBs and often are larger than the storage capacity of single accelerator nodes [122].

Designing a training platform therefore involves building a *scale-out* cluster of multiple accelerators. Examples of such clusters include the Google Cloud TPU [80] that connects multiple TPUs via a 3D torus topology, NVIDIA DGX-2 [123] that connects 16 GPUs using a switch, Facebook's Zion [124] connecting multiple GPUs as a Hypercube Mesh, and so on. The memory system within these platforms needs to provide high capacity and bandwidth. Gradient calculation during distributed training also requires fast transfer of large matrices with minimal latency. Therefore, a high bandwidth network (on-chip/off-chip) needs to be employed to connect the clusters. And finally, similar to the dataflow approaches discussed for inference in this book, training also requires a clever mechanism to map and schedule the training operations— either by distributing the data samples (called data-parallel) or the model (called model-parallel) or leveraging some hybrid approach.

8.4 SUMMARY

The design of DNN accelerators is an extremely nascent field and this book is an attempt to classify and understand various DNN accelerators under a common taxonomy and core design principles. Although the descriptions and examples in this book are limited by workloads and architectures that exist today, we believe many of the core design principles we presented will persist and be useful for computer architects to navigate disruptive innovations across AI workloads and develop efficient future-proof acceleration platforms. In future, we can expect a convergence on some key accelerator architectures once the field of AI matures, target workloads stabilize, and software stacks become robust.

Bibliography

[1] A. Vaswani, N. Shazeer, N. Parmar, J. Uszkoreit, L. Jones, A. N. Gomez, Ł. Kaiser, and I. Polosukhin, Attention is all you need, in *Advances in Neural Information Processing Systems*, pages 5998–6008, 2017. 4, 6

[2] A. Krizhevsky, I. Sutskever, and G. E. Hinton, ImageNet classification with deep convolutional neural networks, in *NIPS*, pages 1097–1105, 2012. DOI: 10.1145/3065386. 4, 5

[3] K. Simonyan and A. Zisserman, Very deep convolutional networks for large-scale image recognition, May 2015. https://arxiv.org/abs/1409.1556 4, 5, 55, 65

[4] C. Szegedy, W. Liu, Y. Jia, P. Sermanet, S. Reed, D. Anguelov, D. Erhan, V. Vanhoucke, and A. Rabinovich, Going deeper with convolutions, in *Proc. of the IEEE Conference on Computer Vision and Pattern Recognition*, pages 1–9, 2015. DOI: 10.1109/cvpr.2015.7298594. 4, 5

[5] K. He, X. Zhang, S. Ren, and J. Sun, Deep residual learning for image recognition, in *Proc. of the IEEE Conference on Computer Vision and Pattern Recognition*, pages 770–778, 2016. DOI: 10.1109/cvpr.2016.90. 4, 5, 8, 26, 83, 89, 92

[6] M. Sandler, A. Howard, M. Zhu, A. Zhmoginov, and L.-C. Chen, MobileNetV2: Inverted residuals and linear bottlenecks, *ArXiv Preprint ArXiv:1801.04381*, 2019. DOI: 10.1109/cvpr.2018.00474. 4, 5

[7] J. Redmon, S. Divvala, R. Girshick, and A. Farhadi, You only look once: Unified, real-time object detection, in *Proc. of the IEEE Conference on Computer Vision and Pattern Recognition*, pages 779–788, 2016. DOI: 10.1109/cvpr.2016.91. 5, 78

[8] O. Ronneberger, P. Fischer, and T. Brox, U-Net: Convolutional networks for biomedical image segmentation, in *International Conference on Medical Image Computing and Computer-Assisted Intervention*, pages 234–241, Springer 2015. DOI: 10.1007/978-3-319-24574-4_28. 5

[9] D. Amodei, S. Ananthanarayanan, R. Anubhai, J. Bai, E. Battenberg, C. Case, J. Casper, B. Catanzaro, Q. Cheng, G. Chen, and J. Chen. Deep speech 2: End-to-end speech recognition in English and Mandarin, in *International Conference on Machine Learning 2016*, pages 173–182, June 11, 2016. 5

[10] Y. Wu, M. Schuster, Z. Chen, Q. V. Le, M. Norouzi, W. Macherey, M. Krikun, Y. Cao, Q. Gao, K. Macherey, and J. Klingner. Google's neural machine translation system: Bridging the gap between human and machine translation, *Preprint ArXiv:1609.08144*, September 26, 2016. 5

[11] A. Vaswani, N. Shazeer, N. Parmar, J. Uszkoreit, L. Jones, A. N. Gomez, Ł. Kaiser, and I. Polosukhin, Attention is all you need, in *Advances in Neural Information Processing Systems*, pages 5998–6008, 2017. 5

[12] A. Radford, J. Wu, R. Child, D. Luan, D. Amodei, and I. Sutskever. Language models are unsupervised multitask learners, *OpenAI Blog*, 1(8):9, February 24, 2019. 5, 79, 126

[13] X. He , L. Liao, H. Zhang, L. Nie, X. Hu, and T. S. Chua, Neural collaborative filtering, in *Proc. of the 26th International Conference on World Wide Web*, pages 173–182, April 3, 2017. DOI: 10.1145/3038912.3052569. 5

[14] M. Naumov, D. Mudigere, H.-J. M. Shi, J. Huang, N. Sundaraman, J. Park, X. Wang, U. Gupta, C.-J. Wu, A. G. Azzolini, et al., Deep learning recommendation model for personalization and recommendation systems, *ArXiv Preprint ArXiv:1906.00091*, 2019. 5

[15] S. Xie, R. Girshick, P. Dollár, Z. Tu, and K. He, Aggregated residual transformations for deep neural networks, in *Proc. of the IEEE Conference on Computer Vision and Pattern Recognition*, pages 1492–1500, 2017. DOI: 10.1109/cvpr.2017.634. 4

[16] M. Tan, B. Chen, R. Pang, V. Vasudevan, M. Sandler, A. Howard, and Q. V. Le, MnasNet: Platform-aware neural architecture search for mobile, in *Proc. of the IEEE Conference on Computer Vision and Pattern Recognition*, pages 2820–2828, 2019. DOI: 10.1109/cvpr.2019.00293. 5

[17] M. Tan and Q. V. Le, EfficientNet: Rethinking model scaling for convolutional neural networks, *ArXiv Preprint ArXiv:1905.11946*, 2019. 5

[18] U. Gupta, C. Wu, X. Wang, M. Naumov, B. Reagen, D. Brooks, B. Cottel, K. M. Hazelwood, M. Hempstead, B. Jia, H. S. Lee, A. Malevich, D. Mudigere, M. Smelyanskiy, L. Xiong, and X. Zhang, The architectural implications of facebook's DNN-based personalized recommendation, in *IEEE International Symposium on High Performance Computer Architecture, (HPCA)*, pages 488–501, San Diego, CA, February 22–26, 2020. DOI: 10.1109/hpca47549.2020.00047. 5

[19] MLPerf. https://github.com/mlperf/reference 5

[20] Y.-H. Chen, J. Emer, and V. Sze, Eyeriss: A spatial architecture for energy-efficient dataflow for convolutional neural networks, in *Proc. of the International Symposium on*

Computer Architecture (ISCA), pages 367–379, June 2016. DOI: 10.1109/isca.2016.40. 9, 11, 13, 15, 27, 28, 33, 34, 41, 47, 48, 53, 56, 66, 71, 88, 97, 111, 114, 115, 126

[21] S. Chetlur, C. Woolley, P. Vandermersch, J. Cohen, J. Tran, B. Catanzaro, and E. Shelhamer, cuDNN: Efficient primitives for deep learning, *ArXiv Preprint ArXiv:1410.0759*, 2014. 29

[22] P. Warden, Why GEMM is at the heart of deep learning, 2015. https://petewarden.com/2015/04/20/why-gemm-is-at-the-heart-of-deep-learning/ 29

[23] W. Qadeer, R. Hameed, O. Shacham, P. Venkatesan, C. Kozyrakis, and M. A. Horowitz, Convolution engine: Balancing efficiency and flexibility in specialized computing, in *ACM SIGARCH Computer Architecture News*, 41:24–35, 2013. DOI: 10.1145/2508148.2485925. 33

[24] Z. Du, R. Fasthuber, T. Chen, P. Ienne, L. Li, T. Luo, X. Feng, Y. Chen, and O. Temam, ShiDianNao: Shifting vision processing closer to the sensor, in *ACM SIGARCH Computer Architecture News*, 43:92–104, 2015. DOI: 10.1145/2872887.2750389. 33, 64, 69

[25] N. P. Jouppi, C. Young, N. Patil, D. Patterson, G. Agrawal, R. Bajwa, S. Bates, S. Bhatia, N. Boden, A. Borchers, R. Boyle, P.-L. Cantin, C. Chao, C. Clark, J. Coriell, M. Daley, M. Dau, J. Dean, B. Gelb, T. V. Ghaemmaghami, R. Gottipati, W. Gulland, R. Hagmann, C. R. Ho, D. Hogberg, J. Hu, R. Hundt, D. Hurt, J. Ibarz, A. Jaffey, A. Jaworski, A. Kaplan, H. Khaitan, D. Killebrew, A. Koch, N. Kumar, S. Lacy, J. Laudon, J. Law, D. Le, C. Leary, Z. Liu, K. Lucke, A. Lundin, G. MacKean, A. Maggiore, M. Mahony, K. Miller, R. Nagarajan, R. Narayanaswami, R. Ni, K. Nix, T. Norrie, M. Omernick, N. Penukonda, A. Phelps, J. Ross, M. Ross, A. Salek, E. Samadiani, C. Severn, G. Sizikov, M. Snelham, J. Souter, D. Steinberg, A. Swing, M. Tan, G. Thorson, B. Tian, H. Toma, E. Tuttle, V. Vasudevan, R. Walter, W. Wang, E. Wilcox, and D. H. Yoon, In-datacenter performance analysis of a tensor processing unit, in *Proc. of the International Symposium on Computer Architecture (ISCA)*, pages 1–12, June 2017. 33, 34, 68, 69, 72, 74, 80, 81, 92, 94, 126, 129

[26] NVIDIA Deep Learning Accelerator (NVDLA), 2017. http://nvdla.org 33, 70, 74, 80, 81, 92, 94

[27] A. Parashar, M. Rhu, A. Mukkara, A. Puglielli, R. Venkatesan, B. Khailany, J. Emer, S. W. Keckler, and W. J. Dally, SCNN: An accelerator for compressed-sparse convolutional neural networks, in *Proc. of the International Symposium on Computer Architecture (ISCA)*, pages 27–40, June 2017. DOI: 10.1145/3079856.3080254. 33, 34, 56, 112, 115, 120

[28] S. Han, X. Liu, H. Mao, J. Pu, A. Pedram, M. A. Horowitz, and W. J. Dally, EIE: Efficient inference engine on compressed deep neural network, in *Proc. of the Inter-*

national Symposium on Computer Architecture (ISCA), pages 243–254, June 2016. DOI: 10.1109/isca.2016.30. 33, 34, 115, 119

[29] D. Liu, T. Chen, S. Liu, J. Zhou, S. Zhou, O. Teman, X. Feng, X. Zhou, and Y. Chen, PuDianNao: A polyvalent machine learning accelerator, in *Proc. of the International Conference on Architectural Support for Programming Languages and Operation Systems (ASPLOS)*, pages 369–381, March 2015. DOI: 10.1145/2694344.2694358. 33, 34

[30] W. J. Dally, Domain specific accelerators. *Keynote at the 52nd International Symposium on Microarchitecture (MICRO)*, 2019. 33, 34

[31] Y. Chen, T. Luo, S. Liu, S. Zhang, L. He, J. Wang, L. Li, T. Chen, Z. Xu, N. Sun, and O. Temam, DaDianNao: A machine-learning supercomputer, in *Proc. of the International Symposium on Microarchitecture (MICRO)*, pages 609–622, December 2014. DOI: 10.1109/micro.2014.58. 34, 56

[32] M. Pellauer, Y. S. Shao, J. Clemons, N. Crago, K. Hegde, R. Venkatesan, S. W. Keckler, C. W. Fletcher, and J. Emer, Buffets: An efficient and composable storage idiom for explicit decoupled data orchestration, in *Architectural Support for Programming Languages and Operating Systems, (ASPLOS)*, 2019. DOI: 10.1145/3297858.3304025. 34, 82

[33] J. Nickolls and W. J. Dally, The GPU computing era, *IEEE Micro*, 30:56–69, March 2010. DOI: 10.1109/mm.2010.41. 38

[34] E. G. Cota, P. Mantovani, G. D. Guglielmo, and L. P. Carloni, An analysis of accelerator coupling in heterogeneous architectures, in *52nd ACM/EDAC/IEEE Design Automation Conference (DAC)*, pages 1–6, June 2015. DOI: 10.1145/2744769.2744794. 39

[35] J. Cong, M. A. Ghodrat, M. Gill, B. Grigorian, K. Gururaj, and G. Reinman, Accelerator-rich architectures: Opportunities and progresses, in *Proc. of the Design Automation Conference (DAC)*, 2014. DOI: 10.1145/2593069.2596667. 39

[36] J. E. Smith, Decoupled access/execute computer architectures, in *Proc. of the International Symposium on Computer Architecture (ISCA)*, pages 112–119, April 1982. DOI: 10.1145/1067649.801719. 40

[37] Xilinx, *FIFO Generator v13.1: LogiCORE IP Product Guide, Vivado Design Suite*, 2017. 40, 44

[38] Intel, *FIFO: Intel FPGA IP User Guide*, 2018. 40, 44

[39] T. Chen and G. E. Suh, Efficient data supply for hardware accelerators with prefetching and access/execute decoupling, in *The Annual IEEE/ACM International Symposium on Microarchitecture (MICRO)*, 2016. DOI: 10.1109/micro.2016.7783749. 41, 57

[40] L. Wu, A. Lottarini, T. Paine, M. Kim, and K. Ross, Q100: The architecture and design of a database processing unit, in *Proc. of the International Conference on Architectural Support for Programming Languages and Operation Systems (ASPLOS)*, 2014. DOI: 10.1145/2541940.2541961. 41

[41] C. F. Fajardo, Z. Fang, R. Iyer, G. F. Garcia, S. E. Lee, and L. Zhao, Buffer-integrated-cache: A cost-effective SRAM architecture for handheld and embedded platforms, in *Proc. of the Design Automation Conference (DAC)*, 2011. DOI: 10.1145/2024724.2024938. 41

[42] W. Lu, G. Yan, J. Li, S. Gong, Y. Han, and X. Li, Flexflow: A flexible dataflow accelerator architecture for convolutional neural networks, in *The International Symposium on High-Performance Computer Architecture (HPCA)*, 2017. DOI: 10.1109/hpca.2017.29. 41, 59, 81

[43] Mentor Graphics, *Catapult Synthesis User and Reference Manual*, 2016. 41

[44] Cadence, *Stratus High-Level Synthesis Reference Guide*, 2015. 41

[45] J. C. Hoe and Arvind, Synthesis of operation-centric hardware descriptions, in *Proc. of the IEEE/ACM International Conference on Computer-Aided Design, (ICCAD)*, pages 511–519, Piscataway, NJ, Press, 2000. DOI: 10.1109/iccad.2000.896524. 41

[46] H. Sharma, J. Park, D. Mahajan, E. Amaro, J. K. Kim, C. Shao, A. Mishra, and H. Esmaeilzadeh, From high-level deep neural models to FPAS, in *49th Annual IEEE/ACM International Symposium on Microarchitecture (MICRO)*, pages 1–12, October 2016. DOI: 10.1109/micro.2016.7783720. 50, 64, 97

[47] Y. Guan, H. Liang, N. Xu, W. Wang, S. Shi, X. Chen, G. Sun, W. Zhang, and J. Cong, FP-DNN: An automated framework for mapping deep neural networks onto FPGAs with RTL-HLS hybrid templates, *IEEE 25th Annual International Symposium on Field-Programmable Custom Computing Machines (FCCM)*, pages 152–159, 2017. DOI: 10.1109/fccm.2017.25. 50

[48] A. Yazdanbakhsh, K. Samadi, N. S. Kim, and H. Esmaeilzadeh, GANAX: A unified MIMD-SIMD acceleration for generative adversarial networks, in *The International Symposium on Computer Architecture (ISCA)*, 2018. 56

[49] J. Fowers, K. Ovtcharov, M. Papamichael, T. Massengill, M. Liu, D. Lo, S. Alkalay, M. Haselman, L. Adams, M. Ghandi, S. Heil, P. Patel, A. Sapek, G. Weisz, L. Woods, S. Lanka, S. K. Reinhardt, A. M. Caulfield, E. S. Chung, and D. Burger, A configurable cloud-scale DNN processor for real-time AI, in *The International Symposium on Computer Architecture (ISCA)*, 2018. DOI: 10.1109/isca.2018.00012. 56

[50] P.-A. Tsai, N. Beckmann, and D. Sanchez, Jenga: Software-defined cache hierarchies, in *Proc. of the International Symposium on Computer Architecture (ISCA)*, 2017. DOI: 10.1145/3079856.3080214. 56, 57

[51] T. J. Ham, J. L. Aragón, and M. Martonosi, DeSC: Decoupled supply-compute communication management for heterogeneous architectures, in *Proc. of the International Symposium on Microarchitecture (MICRO)*, pages 191–203, December 2015. DOI: 10.1145/2830772.2830800. 57

[52] R. Komuravelli, M. D. Sinclair, J. Alsop, M. Huzaifa, M. Kotsifakou, P. Srivastava, S. V. Adve, and V. S. Adve, Stash: Have your scratchpad and cache it too, in *Proc. of the International Symposium on Computer Architecture (ISCA)*, pages 707–719, June 2015. DOI: 10.1145/2749469.2750374. 57

[53] M. J. Lyons, M. Hempstead, G.-Y. Wei, and D. Brooks, The accelerator store: A shared memory framework for accelerator-based systems, *ACM Transactions on Architecture and Code Optimization*, 8:48:1–48:22, January 2012. DOI: 10.1145/2086696.2086727. 57

[54] J. Clemons, C. C. Cheng, I. Frosio, D. Johnson, and S. W. Keckler, A patch memory system for image processing and computer vision, in *Proc. of the International Symposium on Microarchitecture (MICRO)*, pages 1–13, October 2016. DOI: 10.1109/micro.2016.7783754. 57, 58

[55] M. Adler, K. E. Fleming, A. Parashar, M. Pellauer, and J. Emer, Leap scratchpads: Automatic memory and cache management for reconfigurable logic, in *Proc. of the International Symposium on Field Programmable Gate Arrays (FPGA)*, pages 25–28, February 2011. DOI: 10.1145/1950413.1950421. 57, 58

[56] E. S. Chung, J. C. Hoe, and K. Mai, CoRAM: An in-fabric memory architecture for FPGA-based computing, in *Proc. of the International Symposium on Field Programmable Gate Arrays (FPGA)*, pages 97–106, February 2011. DOI: 10.1145/1950413.1950435. 57, 58

[57] T. Nowatzki, V. Gangadhar, N. Ardalani, and K. Sankaralingam, Stream-dataflow acceleration, in *Proc. of the International Symposium on Computer Architecture (ISCA)*, 2017. DOI: 10.1145/3079856.3080255. 57, 58

[58] H. Kwon, A. Samajdar, and T. Krishna, MAERI: Enabling flexible dataflow mapping over DNN accelerators via reconfigurable interconnects, in *International Conference on Architectural Support for Programming Languages and Operating Systems, (ASPLOS)*, pages 461–475, 2018. DOI: 10.1145/3173162.3173176. 59, 68, 69, 70, 74, 81, 88, 89, 90, 92, 94, 97

[59] L. Song, Y. Wang, Y. Han, X. Zhao, B. Liu, and X. Li, C-Brain: A deep learning accelerator that tames the diversity of CNNs through adaptive data-level parallelization, in *Proc. of the Design Automation Conference (DAC)*, 2016. DOI: 10.1145/2897937.2897995. 59

[60] C. Farabet, B. Martini, B. Corda, P. Akselrod, E. Culurciello, and Y. LeCun, Neuflow: A runtime reconfigurable dataflow processor for vision, in *Computer Vision and Pattern Recognition Workshops (CVPRW)*, 2011. DOI: 10.1109/cvprw.2011.5981829. 59

[61] Y. S. Shao, J. Clemons, R. Venkatesan, B. Zimmer, M. Fojtik, N. Jiang, B. Keller, A. Klinefelter, N. R. Pinckney, P. Raina, S. G. Tell, Y. Zhang, W. J. Dally, J. S. Emer, C. T. Gray, B. Khailany, and S. W. Keckler, Simba: Scaling deep-learning inference with multi-chip-module-based architecture, in *Proc. of the 52nd Annual IEEE/ACM International Symposium on Microarchitecture, (MICRO)*, pages 12–16, Columbus, OH, October 2019. DOI: 10.1145/3352460.3358302. 61, 64, 82, 92, 96, 126

[62] N. Enright Jerger, T. Krishna, and L.-S. Peh, *On-Chip Networks*, 2nd ed., Synthesis Lectures on Computer Architecture, Morgan & Claypool, 2017. DOI: 10.2200/s00772ed1v01y201704cac040. 62, 64

[63] S. Murali and G. De Micheli, SUNMAP: A tool for automatic topology selection and generation for NoCs, in *Proc. of the 41st Annual Design Automation Conference*, pages 914–919, 2004. DOI: 10.1145/996566.996809. 62

[64] Y.-H. Chen, T. Krishna, J. Emer, and V. Sze, Eyeriss: An energy-efficient reconfigurable accelerator for deep convolutional neural networks, in *Proc. of the International Solid State Circuits Conference (ISSCC)*, February 2016. DOI: 10.1109/isscc.2016.7418007. 63, 92, 111

[65] T. Chen, Z. Du, N. Sun, J. Wang, C. Wu, Y. Chen, and O. Temam, DianNao: A small-footprint high-throughput accelerator for ubiquitous machine-learning, in *Proc. of the 19th International Conference on Architectural Support for Programming Languages and Operating Systems*, pages 269–284, 2014. DOI: 10.1145/2654822.2541967. 64

[66] F. Akopyan, J. Sawada, A. Cassidy, R. Alvarez-Icaza, J. Arthur, P. Merolla, N. Imam, Y. Nakamura, P. Datta, G.-J. Nam, et al., TrueNorth: Design and tool flow of a 65 mw 1 million neuron programmable neurosynaptic chip, *IEEE Transactions on Computer-Aided Design of Integrated Circuits and Systems*, 34(10):1537–1557, 2015. DOI: 10.1109/tcad.2015.2474396. 64

[67] W. J. Dally and B. P. Towles, *Principles and Practices of Interconnection Networks*, Elsevier 2004. 64

[68] H. Kwon, A. Samajdar, and T. Krishna, Rethinking NoCs for spatial neural network accelerators, in *11th IEEE/ACM International Symposium on Networks-on-Chip (NOCS)*, pages 1–8, 2017. DOI: 10.1145/3130218.3130230. 66, 68, 72

[69] Y.-H. Chen, T.-J. Yang, J. Emer, and V. Sze, Eyeriss v2: A flexible accelerator for emerging deep neural networks on mobile devices, *IEEE Journal on Emerging and Selected Topics in Circuits and Systems*, 2019. DOI: 10.1109/jetcas.2019.2910232. 68, 72, 92, 95, 114

[70] E. Chung, J. Fowers, K. Ovtcharov, M. Papamichael, A. Caulfield, T. Massengill, M. Liu, D. Lo, S. Alkalay, M. Haselman, et al., Serving DNNs in real time at datacenter scale with project brainwave, *IEEE Micro*, 38(2):8–20, 2018. DOI: 10.1109/mm.2018.022071131. 70, 82

[71] H. Kwon, L. Lai, T. Krishna, and V. Chandra, Herald: Optimizing heterogeneous DNN accelerators for edge devices, *ArXiv Preprint ArXiv:1909.07437*, 2019. 78

[72] M. Shoeybi, M. Patwary, R. Puri, P. LeGresley, J. Casper, and B. Catanzaro, Megatron-LM: Training multi-billion parameter language models using GPU model parallelism, *ArXiv Preprint ArXiv:1909.08053*, 2019. 79, 126

[73] Y. Gan, Y. Zhang, D. Cheng, A. Shetty, P. Rathi, N. Katarki, A. Bruno, J. Hu, B. Ritchken, B. Jackson, et al., An open-source benchmark suite for microservices and their hardware-software implications for cloud and edge systems, in *Proc. of the 24th International Conference on Architectural Support for Programming Languages and Operating Systems*, pages 3–18, 2019. DOI: 10.1145/3297858.3304013. 79

[74] M. Gao, J. Pu, X. Yang, M. Horowitz, and C. Kozyrakis, Tetris: Scalable and efficient neural network acceleration with: 3D memory, in *Proc. of the 22nd International Conference on Architectural Support for Programming Languages and Operating Systems, (ASPLOS)*, pages 751–764, ACM, New York, 2017. DOI: 10.1145/3037697.3037702. 82, 97, 117, 126

[75] M. Gao, X. Yang, J. Pu, M. Horowitz, and C. Kozyrakis, Tangram: Optimized coarse-grained dataflow for scalable NN accelerators, in *Proc. of the 24th International Conference on Architectural Support for Programming Languages and Operating Systems*, pages 807–820, ACM, 2019. DOI: 10.1145/3297858.3304014. 82

[76] E. Qin, A. Samajdar, H. Kwon, V. Nadella, D. Das, S. Srinivasan, B. Kaul, and T. Krishna, SIGMA: A sparse and irregular GEMM accelerator with flexible interconnects for DNN training, in *IEEE International Symposium on High Performance Computer Architecture*, 2020. DOI: 10.1109/hpca47549.2020.00015. 82, 125

[77] B. Jacob, S. Kligys, B. Chen, M. Zhu, M. Tang, A. Howard, H. Adam, and D. Kalenichenko, Quantization and training of neural networks for efficient integer-arithmetic-only inference, in *Proc. of the IEEE Conference on Computer Vision and Pattern Recognition*, pages 2704–2713, 2018. DOI: 10.1109/cvpr.2018.00286. 85

[78] A. Samajdar, J. M. Joseph, Y. Zhu, P. Whatmough, M. Mattina, and T. Krishna, A systematic methodology for characterizing scalability of DNN accelerators using SCALE-Sim, in *IEEE International Symposium on Performance Analysis of Systems and Software*, 2020. 87, 88

[79] H. Kwon, P. Chatarasi, M. Pellauer, A. Parashar, V. Sarkar, and T. Krishna, Understanding reuse, performance, and hardware cost of DNN dataflows: A data-centric approach, in *Proc. of the 52nd Annual IEEE/ACM International Symposium on Microarchitecture, (MICRO)*, pages 754–768, Columbus, OH, October 12–16, 2019. DOI: 10.1145/3352460.3358252. 92, 101, 105, 106

[80] Tearing Apart Google's TPU:3.0 AI Co Processor, 2018. https://www.nextplatform.com/2018/05/10/tearing-apart-googles-tpu-3-0-ai-coprocessor/ 94, 129

[81] Y.-H. Chen, J. Emer, and V. Sze, Using dataflow to optimize energy efficiency of deep neural network accelerators, *IEEE Micro's Top Picks from the Computer Architecture Conferences*, 37, May–June 2017. DOI: 10.1109/mm.2017.54. 97

[82] A. Rahman, S. Oh, J. Lee, and K. Choi, Design space exploration of FPGA accelerators for convolutional neural networks, in *Design, Automation and Test in Europe Conference and Exhibition (DATE)*, 2017. DOI: 10.23919/date.2017.7927162. 97

[83] M. Motamedi, P. Gysel, V. Akella, and S. Ghiasi, Design space exploration of FPGA-based deep convolutional neural networks, in *21st Asia and South Pacific Design Automation Conference (ASP-DAC)*, 2016. DOI: 10.1109/aspdac.2016.7428073. 97

[84] K. Yang, S. Wang, J. Zhou, and T. Yoshimura, Energy-efficient scheduling method with cross-loop model for resource-limited CNN accelerator designs, in *IEEE International Symposium on Circuits and Systems (ISCAS)*, 2017. DOI: 10.1109/iscas.2017.8050800. 97

[85] K. Ueyoshi, K. Ando, K. Orimo, M. Ikebe, T. Asai, and M. Motomura, Exploring optimized accelerator design for binarized convolutional neural networks, in *International Joint Conference on Neural Networks (IJCNN)*, 2017. DOI: 10.1109/ijcnn.2017.7966161. 97

[86] Y. Shen, M. Ferdman, and P. Milder, Maximizing CNN accelerator efficiency through resource partitioning, in *Proc. of the International Symposium on Computer Architecture (ISCA)*, 2017. DOI: 10.1145/3079856.3080221. 97

[87] Y. Ma, Y. Cao, S. Vrudhula, and J.-S. Seo, Optimizing loop operation and dataflow in FPGA acceleration of deep convolutional neural networks, in *Proc. of the ACM/SIGDA International Symposium on Field-Programmable Gate Arrays*, 2017. DOI: 10.1145/3020078.3021736. 97

[88] A. Parashar, P. Raina, Y. Sophia Shao, Y.-H. Chen, V. A. Ying, A. Mukkara, R. Venkatesan, B. Khailany, S. W. Keckler, and J. Emer, Timeloop: A systematic approach to DNN accelerator evaluation, in *IEEE International Symposium on Performance Analysis of Systems and Software, (ISPASS)*, pages 304–315, March 2019. DOI: 10.1109/ispass.2019.00042. 97, 98, 100, 105, 106, 107, 108, 109

[89] T.-J. Yang, Y.-H. Chen, J. Emer, and V. Sze, A method to estimate the energy consumption of deep neural networks, in *Asilomar Conference on Signals, Systems and Computers*, 2017. DOI: 10.1109/acssc.2017.8335698. 97

[90] L. Ke, X. He, and X. Zhang, NNest: Early-stage design space exploration tool for neural network inference accelerators, in *Proc. of the International Symposium on Low Power Electronics and Design, (ISLPED)*, pages 4:1–4:6, ACM, New York, 2018. DOI: 10.1145/3218603.3218647. 97

[91] Y. N. Wu, J. S. Emer, and V. Sze, Accelergy: An architecture-level energy estimation methodology for accelerator designs, in *IEEE/ACM International Conference On Computer Aided Design (ICCAD)*, 2019. DOI: 10.1109/iccad45719.2019.8942149. 108

[92] N. Vasilache, O. Zinenko, T. Theodoridis, P. Goyal, Z. DeVito, W. S. Moses, S. Verdoolaege, A. Adams, and A. Cohen, Tensor comprehensions: Framework-agnostic high-performance machine learning abstractions, *CoRR*, 2018. 109

[93] R. Baghdadi, J. Ray, M. B. Romdhane, E. D. Sozzo, A. Akkas, Y. Zhang, P. Suriana, S. Kamil, and S. P. Amarasinghe, Tiramisu: A polyhedral compiler for expressing fast and portable code, *CoRR*, 2018. DOI: 10.1109/cgo.2019.8661197. 109

[94] A. Adams, K. Ma, L. Anderson, R. Baghdadi, T.-M. Li, M. Gharbi, B. Steiner, S. Johnson, K. Fatahalian, F. Durand, and J. Ragan-Kelley, Learning to optimize halide with tree search and random programs, *ACM Transactions on Graphic*, 38, July 2019. DOI: 10.1145/3306346.3322967. 109

[95] B. H. Ahn, P. Pilligundla, and H. Esmaeilzadeh, Reinforcement learning and adaptive sampling for optimized DNN compilation, *CoRR*, 2019. 109

[96] T. Chen, T. Moreau, Z. Jiang, L. Zheng, E. Yan, H. Shen, M. Cowan, L. Wang, Y. Hu, L. Ceze, C. Guestrin, and A. Krishnamurthy, TVM: An automated end-to-end optimizing compiler for deep learning, in *13th USENIX Symposium on Operating Systems*

Design and Implementation (OSDI), pages 578–594, USENIX Association, Carlsbad, CA, 2018. 109

[97] S. Dave, Y. Kim, S. Avancha, K. Lee, and A. Shrivastava, dMazeRunner: Executing perfectly nested loops on dataflow accelerators, *ACM Transactions on Embedded Computing Systems*, 18, October 2019. DOI: 10.1145/3358198. 109

[98] S. Han, J. Pool, J. Tran, and W. Dally, Learning both weights and connections for efficient neural networks, in *Proc. of the International Conference on Neural Information Processing Systems (NIPS)*, pages 1135–1143, December 2015. 111, 112, 124

[99] S. Han, J. Pool, J. Tran, and W. Dally, Deep compression: Compressing deep neural networks with pruning, trained quantization and Huffman coding, in *arxiv.org*, 2015. 111

[100] J. Albericio, P. Judd, T. Hetherington, T. Aamodt, N. Enright Jerger, and A. Moshovos, Cnvlutin: Ineffectual-neuron-free deep convolutional neural network computing, in *Proc. of the International Symposium on Computer Architecture (ISCA)*, pages 1–13, June 2016. DOI: 10.1109/isca.2016.11. 112, 114, 119

[101] S. Zhang, Z. Du, L. Zhang, H. Lan, S. Liu, L. Li, Q. Guo, T. Chen, and Y. Chen, Cambricon-X: An accelerator for sparse neural networks, in *Proc. of the International Symposium on Microarchitecture (MICRO)*, October 2016. DOI: 10.1109/micro.2016.7783723. 112, 114, 119

[102] C. Lee, Y. Shao, J.-F. Zhang, A. Parashar, J. Emer, S. Keckler, and Z. Zhang, Stitch-X: An accelerator architecture for exploiting unstructured sparsity in deep neural networks, in *SysML Conference*, 2018. 120

[103] A. Zhou, A. Yao, Y. Guo, L. Xu, and Y. Chen, Incremental network quantization: Towards lossless CNNs with low-precision weights, in *ICLR*, 2017. 124

[104] C. Zhu, S. Han, H. Mao, and W. J. Dally, Trained ternary quantization, *ArXiv Preprint ArXiv:1612.01064*, 2016. 124

[105] J. Qiu, J. Wang, S. Yao, K. Guo, B. Li, E. Zhou, J. Yu, T. Tang, N. Xu, S. Song, Y. Wang, and H. Yang, Going deeper with embedded FPGA platform for convolutional neural network, in *Proc. of the ACM/SIGDA International Symposium on Field-Programmable Gate Arrays, (FPGA)*, pages 26–35, Association for Computing Machinery, New York, 2016. DOI: 10.1145/2847263.2847265. 124

[106] J. Lee, D. Shin, and H. Yoo, A 21 mw low-power recurrent neural network accelerator with quantization tables for embedded deep learning applications, in *IEEE Asian Solid-State Circuits Conference (A-SSCC)*, pages 237–240, 2017. DOI: 10.1109/asscc.2017.8240260. 124

[107] S. Han, J. Pool, S. Narang, H. Mao, E. Gong, S. Tang, E. Elsen, P. Vajda, M. Paluri, J. Tran, B. Catanzaro, and W. J. Dally, DSD: Dense-sparse-dense training for deep neural networks, *ArXiv: Computer Vision and Pattern Recognition*, 2017. 125

[108] S. Pal, J. Beaumont, D.-H. Park, A. Amarnath, S. Feng, C. Chakrabarti, H.-S. Kim, D. Blaauw, T. Mudge, and R. Dreslinski, Outerspace: An outer product based sparse matrix multiplication accelerator, in *IEEE International Symposium on High Performance Computer Architecture (HPCA)*, pages 724–736, 2018. DOI: 10.1109/hpca.2018.00067. 125

[109] K. Hegde, H. Asghari-Moghaddam, M. Pellauer, N. Crago, A. Jaleel, E. Solomonik, J. Emer, and C. W. Fletcher, Extensor: An accelerator for sparse tensor algebra, in *Proc. of the 52nd Annual IEEE/ACM International Symposium on Microarchitecture, (MICRO)*, pages 319–333, Association for Computing Machinery, New York, 2019. DOI: 10.1145/3352460.3358275. 125

[110] Z. Zhang, H. Wang, S. Han, and W. J. Dally, Sparch: Efficient architecture for sparse matrix multiplication, in *IEEE International Symposium on High Performance Computer Architecture (HPCA)*, pages 261–274, 2020. DOI: 10.1109/hpca47549.2020.00030. 125

[111] S. Winograd, Complexity of computations, in *Proc. of the Annual Conference, (ACM)*, pages 138–141, Association for Computing Machinery, New York, 1978. DOI: 10.1145/800127.804083. 125

[112] B. Barabasz and D. Gregg, Winograd convolution for DNNs: Beyond linear polynomials, in *International Conference of the Italian Association for Artificial Intelligence*, pages 307–320, Springer, Cham, 2019. DOI: 10.1007/978-3-030-35166-3_22. 125

[113] K. Hegde, J. Yu, R. Agrawal, M. Yan, M. Pellauer, and C. W. Fletcher, UCNN: Exploiting computational reuse in deep neural networks via weight repetition, in *Proc. of the 45th Annual International Symposium on Computer Architecture, (ISCA)*, pages 674–687, IEEE Press, 2018. DOI: 10.1109/isca.2018.00062. 125

[114] H. Sharma, J. Park, N. Suda, L. Lai, B. Chau, V. Chandra, and H. Esmaeilzadeh, Bit fusion: Bit-level dynamically composable architecture for accelerating deep neural networks, in *Proc. of the 45th Annual International Symposium on Computer Architecture, (ISCA)*, pages 764–775, IEEE Press, 2018. DOI: 10.1109/isca.2018.00069. 125

[115] P. Judd, J. Albericio, T. Hetherington, T. M. Aamodt, and A. Moshovos, Stripes: Bit-serial deep neural network computing, in *49th Annual IEEE/ACM International Symposium on Microarchitecture (MICRO)*, pages 1–12, 2016. DOI: 10.1109/micro.2016.7783722. 125

[116] Cerebras Wafer Scale Engine: Why we need big chips for Deep Learning, 2019. https://www.cerebras.net/cerebras-wafer-scale-engine-why-we-need-big-chips-for-deep-learning/ 126

[117] W. Haensch, T. Gokmen, and R. Puri, The next generation of deep learning hardware: Analog computing, in *Proc. of the IEEE*, 107(1):108–122, 2018. DOI: 10.1109/jproc.2018.2871057. 127

[118] Xilinx, Xilinx ML Suite, 2019. https://github.com/Xilinx/ml-suite 128

[119] Xilinx, Xilinx Versal AI Core, 2019. https://www.xilinx.com/products/silicon-devices/acap/versal-ai-core.html 128

[120] NVIDIA, NVIDIA Volta, 2017. https://www.nvidia.com/en-us/data-center/volta-gpu-architecture/ 128

[121] R. Krashinsky, O. Giroux, S. Jones, N. Stam, and S. Ramaswamy, NVIDIA Ampere Architecture In-Depth, 2020. https://devblogs.nvidia.com/nvidia-ampere-architecture-in-depth/ 128

[122] M. Rhu, N. Gimelshein, J. Clemons, A. Zulfiqar, and S. W. Keckler, vDNN: Virtualized deep neural networks for scalable, memory-efficient neural network design, in *Proc. of the International Symposium on Microarchitecture (MICRO)*, October 2016. DOI: 10.1109/micro.2016.7783721. 129

[123] NVIDIA DGX-2, 2018. https://https://www.nvidia.com/content/dam/en-zz/Solutions/Data-Center/dgx-2/dgx-2-print-datasheet-738070-nvidia-a4-web-uk.pdf/ 129

[124] M. Smelyanskiy, Zion: Facebook next-generation large memory training platform, in *IEEE Hot Chips 31 Symposium (HCS)*, pages 1–22, 2019. DOI: 10.1109/hotchips.2019.8875650. 129

Authors' Biographies

TUSHAR KRISHNA

Tushar Krishna is an Assistant Professor in the School of Electrical and Computer Engineering at the Georgia Institute of Technology. He received a Ph.D. in Electrical Engineering and Computer Science from the Massachusetts Institute of Technology in 2014. Prior to that, he received an M.S.E in Electrical Engineering from Princeton University in 2009 and a B.Tech in Electrical Engineering from the Indian Institute of Technology (IIT), Delhi in 2007. Before joining Georgia Tech in 2015, he worked as a researcher in the VSSAD Group at Intel in Massachusetts. Dr. Krishna's research spans computer architecture, interconnection networks, networks-on-chip (NoC), and deep learning accelerators, with a focus on optimizing data movement in modern computing systems. Three of his papers have been selected for IEEE Micro's Top Picks from Computer Architecture, one more received an honorable mention, and three have won best paper awards. He received the National Science Foundation (NSF) CRII award in 2018 and both a Google Faculty Award and a Facebook Faculty Award in 2019.

HYOUKJUN KWON

Hyoukjun Kwon is a research scientist at Facebook AR/VR. He received his Ph.D. in Computer Science from Georgia Institute of Technology in 2020, advised by Dr. Tushar Krishna. He received B.S. degrees in Environmental Materials Science and in Computer Science and Engineering from Seoul National University in 2015. His research interests include communication-centric DNN accelerator designs, modeling of DNN accelerator architecture and mapping, NoC for accelerators, and co-optimization of DNN model, mapping, and accelerator architecture. He is actively leading the development of multiple open-source tools and RTLs in the DNN accelerator domain, including MAESTRO, MAERI, Microswitch NoC, and OpenSMART. One of his papers was selected for IEEE Micro's Top Picks from computer architecture in 2019, one received honorable mention in 2018, and another won the best paper award at HPCA 2020.

ANGSHUMAN PARASHAR

Angshuman Parashar is a Senior Research Scientist at NVIDIA. His research interests are in building, evaluating, and programming spatial and data-parallel architectures, with a present focus on automated mapping of machine learning algorithms onto architectures based on explicit decoupled data orchestration. Prior to NVIDIA, he was a member of the VSSAD group at

Intel, where he worked with a small team of experts in architecture, languages, workloads, and implementation to design and evaluate a new spatial architecture. Dr. Parashar received his Ph.D. in Computer Science and Engineering from the Pennsylvania State University in 2007, and his B.Tech. in Computer Science and Engineering from the Indian Institute of Technology, Delhi in 2002.

MICHAEL PELLAUER

Michael Pellauer is a Senior Research Scientist at NVIDIA. His research interest is building domain specific accelerators, with a special emphasis on deep learning and sparse tensor algebra. Prior to NVIDIA, he was a member of the VSSAD group at Intel, where he performed research and advanced development on customized spatial accelerators. Dr. Pellauer holds a Ph.D. from the Massachusetts Institute of Technology in Cambridge, Massachusetts (2010), a Master's from Chalmers University of Technology in Gothenburg, Sweden (2003), and a Bachelor's from Brown University in Providence, Rhode Island (1999).

ANANDA SAMAJDAR

Ananda Samajdar is a Ph.D. student at the school of Electrical and Computer Engineering (ECE) at the Georgia Institute of Technology. He completed his B.Tech. (Hons.) in Electronics and Communication Engineering (ECE) from the Indian Institute of Information Technology, Allahabad India (IIIT-A) in 2013. Before joining Georgia Tech, Anand worked as a VLSI design engineer at Qualcomm Bangalore for three years. Anand's research interest includes designing custom architecture for efficient and deep learning systems. He has authored a number of papers in top-tier computer architecture conferences. Two of his papers received honorable mentions in the IEEE MICRO Top Picks 2019, and one was awarded the best paper award at HPCA 2020. He is also the recipient of the silver medal for the ACM student research competition at ASPLOS 2019.

Printed in the United States
by Baker & Taylor Publisher Services